Solutions Man

Materials and Processes in Manufacturing

EIGHTH EDITION

DR S·O· ONUOH

E. PAUL DeGARMO
J T. BLACK
RONALD A. KOHSER

PRENTICE HALL, UPPER SADDLE RIVER, NJ 07458

Acquisition Editor: *Bill Stenquist*
Production Editor: *Shea Oakley*
Special Projects Manager: *Barbara A. Murray*
Supplement Cover Manager: *Paul Gourhan*
Manufacturing Buyer: *Donna Sullivan*

Printed in the United States of America

10 9 8 7 6 5 4 3

ISBN 0-02-328623-7

Prentice-Hall International (UK) Limited, *London*
Prentice-Hall of Australia Pty. Limited, *Sydney*
Prentice-Hall Canada Inc., *Toronto*
Prentice Hall Hispanoamericana, S.A., *Mexico*
Prentice-Hall of India Private Limited, *New Delhi*
Prentice-Hall of Japan, Inc., *Tokyo*
Simon & Schuster Asia Pte. Ltd., *Singapore*
Editoria Prentice-Hall do Brasil, *Ltda., Rio de Janero*

Contents

Chapter 1

INTRODUCTION

1. The availability and cost of manufactured products are an important part of our cost of living and the real wealth of the nation. Thus, reducing the cost of producer and consumer goods improves the productivity while holding down inflation, thereby improving the general standard of living.

2. This is true if you consider that everyone who used the output from a process, including all the intermediate steps, is a customer. The operator of the next process is the user and customer of the proceeding process. In fact, some companies identify two customers, the external customer who buys the finished product and the internal customer, who builds the product one - i.e., the people who work in the manufacturing system. See Chapter 43

3. A job shop is a manufacturing system that is designed (or layed out) in a functional manner (all kinds of equipment grouped together) capable of producing great variety with highly skilled labor.

   A flow shop uses more special purpose equipment to produce a specific group of products. They utilize specialized machines and are generally layed out so that the specific products pass through each machine in a series such that no machines are skipped, and no backflow is allowed. The product range is smaller, labor is usually less skilled, and the volume of product is large. The flow shop is usually viewed as "mass production".

   The project shop is a manufacturing system wherein the end item being built is too large to be easily moved about the plant (a locomotive, a large machine or a large plane) or is stationary (a building or bridge). Thus, the men, materials, supplies, equipment, and machine tools must all come to the item, which frequently is a unique, one-of-a-kind design.

   A continuous process is a manufacturing system wherein the end product can flow through the process and it runs continuously or over long periods of time to produce large batches, usually of liquids or chemicals.

   The cellular system is composed of manufacturing cells and subassembly cells linked together directly or with Kanban. Each manufacturing cell is a collection of processes designed to make a family of parts.

4. In the context of manufacturing, a manufacturing system is a collection of men, machine tools, and material-moving systems, collected together to accomplish specific manufacturing or fabrication sequences, resulting in components or end products. The manufacturing system is backed up by and supported by the production system, which includes functions like control of quality, inventory, production, and manpower, as well as scheduling, planning and the like. Within the manufacturing system, there will be machine tools, which can perform jobs or

tasks. A series of operations or tasks are called processes or a sequence of operations. In practice, in industry, these terms are often carelessly used.

5. No, a cutting tool is the thing that actually contacts the metal in a machining operation and produces the chip. Cutting tools are held in tool holders, which are part of the machine tool (which usually includes the bed, motor, frame, etc.) Cutting tools are usually purchased separately from the machine. Tooling usually refers to work holding devices in the machine tools.

6. The basic manufacturing processes are: casting or molding, forming, (heat) treating, metal removal, finishing, assembling, and inspection.

7. By casting, the desired shaped in final or near-final form, could be obtained. This greatly reduces the necessity for machining the hard-to-machine metal. Less machining is needed when the raw material shape is close to the finished part size and shape (called near net shape casting).

8. Trains stop at the station to load and unload people and materials. In an assembly line, products stop at the job station to take on materials or have operations performed on them.

9. False. Storage is very expensive because time costs the company money. It is expensive to keep track of stored materials, to put them into storage, to get them back from storage, to damage them as a result of excessive handling, and so on. More importantly, storage usually adds no value - very few items appreciate on the shelf.

10. The initial processes require that the material, usually a low-stength steel, be formed into bar stock of small enough diameter that it can be converted into wire in a "wire drawing" process, the latter being a series of machines which pull the wire through dies, reducing the wire diameter in steps. The wire is cut to length, bent into a hanger and finally twisted. All of these processes must produce a sufficient volume of hangers at a low enough unit cost that they can be given away by the cleaning establishment.

11. The university is an example of a service job shop and shows that value can be added by service processes and operations --the student enters engineering worth the minimum wage and graduates worth $15 to $20/hour. In the university job shop, the professors are the machine tool operators, the students are the workpieces, courses are the processes, tests are the inspections, books are the tooling, and department heads are the foremen.

12. Inefficient is a relative term here. If we can eliminate machining, we can save the time and the money. Machining processes are generally those which give the part its final size, shape, and surface finish and add value to the part. Because they do not produce the shape and size in bulk, but rather by localized action they may not be as efficient as forming and casting processes.

13. The level of automation as practiced by the surgeon or the plumber is A(0). They both used hand tools almost exclusively and the quality of the finished job is almost totally dependent upon their dexterity, skill and judgment during the operations. Some laser surgery has been automated.

14. Preserve the product, protect the produce, and merchandise the product.

15. Eli Whitney is generally given credit for the invention of filing jigs.

16. To respond to this question, the student should observe some assembly process and list or describe the steps using the language of manufacturing. Making a sandwich can be considered as an assembly process, and in fact, the airlines mass assembly sandwiches (snacks) for their customers. Building a house is not considered assembly, but rather project fabrication. However, there are house manufactures who fabricate houses on assembly lines and truck them to the building site for final assembly.

17. A slaughter house is a "disassembly" factory for cows or pigs. The whole animal is divided into parts. It is said that Henry Ford got the idea for the assembly line while watching animals being slaughtered.

18. This is really a discussion question to get the students to be aware of all the things involved in characterizing a process technology. The extrusion process results when the pressure applied to a material exceeds its flow strength. Sufficient energy must be applied to overcome friction, so lubrication is very important. The tooling is generally very expensive. A single die may cost $5000 and setup time can be long. The process usually produces 10 to 25 surface feet per minute of material. The critical process parameters are temperature and pressure, the material being extruded, lubrication, and extrusion rate. Some metals cannot be extruded very well. The process is constrained by the power available and the size of the billets -- i.e. the standard process is not continuous. The process operates reliably but users should always be aware of the high pressures involved in upsetting the materials. Operator skills are not critical and the process is semiautomatic. The process can do a wide variety of parts, depending only on the die design. It is hard to do hollow extrusions. Extrusion as a process is typically good to a tolerance of about 0.001 or 0.002 inches.

19. Mechanization involves the energizing of the machines and providing them with stock or material feeding devices. They typically require a human operator. Automation includes mechanization and provides for repeated cycles and self-feeding at level A(3) and feedback and corrective actions at level A(4) and above. See the Yardstick for Automation in Chapters 1 and 29.

20. An example of an A(0) machine would be a push-type lawn mower, while a gas powered mower would be an example of an A(1) machine. A single cycle automatic would be a toaster, an electric dishwasher, or a modern washing machine. A multiple-disc CD-player that can change discs would be an example of an A(3) machine, which processes CDs (i.e. plays them). Controls of the type found in A(4) machines are used in controlling the temperature of a house or the oven in a kitchen. The water level in the toilet flush tank uses feedback as well.

21. Production planning is deciding <u>what</u> should be done and how it should be done, what machines should be used, in what sequences, to make a part, and how these machines should be tooled, set-up, and operated. Scheduling is deciding <u>when</u> the production should take place, and therefore, when parts and products should be completed and ready for sale.   Without these kinds of critical functions in the production system, the manufacturing system would grind to an inefficient halt.

22. How would a bumper have to be redesigned to provide the equivalent strength? What other components would have to be redesigned? What additional or different processing equipment, including finishing equipment, would be needed? Would the aluminum bumper satisfy the safety requirements (5 mph crash test) needed by the car? What are the costs savings produced by this change?

23. To a great degree, the design of the part dictates the possible ways it can be fabricated. The determination of the material(s) and the quantity to be made are also critical factors. The design can make manufacturing and assembly easier and more economical or difficult and costly. This effort on the part of the product designer is called design for manufacturing and assembly. It is even more important when the product (or component) is being designed for automatic or robotic assembly. See additional references on automation.

24. It is almost impossible to fabricate a low-cost item that is poorly designed and do it in a economical way. It must be designed so that it is easy to produce if it is going to be inexpensive (i.e. it has producibility). Thus, this statement is true.

25. Operations like load and unload parts from the machine, change the machine over from one part design to another (this is called set up), and change the tooling in the machine all add no value.

4

26. A proprietary process is one which the company tries to keep secret from their competitors. They believe that they know something about the process that their competition does not know. This is referred to as uniqueness in the manufacturing processes or system and requires creativity on the part of the engineer to develop new processes to make products which the competition can not duplicate. Such processes are not patented because that would force the company to reveal how they worked.

27. The electric typewriter is A(1).

28. The rolls produce many feet of sheet metal that end up in many cars, so the fixed costs (like the rolls) are spread out over many sheets (feet) of metal. Thus the cost of sheet metal per car may be 50 to 100 dollars before the metal is formed into fenders and door panels.

29. Insurance, health, entertainment, sporting events, transportation, lodging, banking, communications, education, etc. are examples of service industries -- anything bought or sold in trade that cannot be dropped on your foot. Service industries worry about productivity, quality, and economic output just as much as manufacturing industries.

30. Disassembling it adds costs and value - you want the cuts of meat, not the whole animal. You are adding value to the cow when you are raising it and feeding it, so it becomes more valuable in the market. You add cost, not value when you ship the cow to the market.

31. Chrome plating is a finishing process.

32. The selling price is determined by the marketplace and what the customer will pay. The best way to improve profit is to reduce manufacturing costs per unit. This can be difficult to do when the price keeps going up.

33. The manufacturing cost for an assembled product like a car is made up of materials (raw materials, cutting tools, purchased parts, and components, and their storage and handling), direct labor, indirect labor (people who work in the manufacturing system but don't work directly on the car), energy, and depreciation (of the machines and tooling.)

PROBLEMS FOR CHAPTER 1

1. If we are to assume that 10% of the total price of the $8000 car goes for direct labor, then $800 is spent on direct labor. Another way to estimate this is from the number of direct labor hours that goes into the car. For American cars, this is about 40 hours depending on the car and the company. If you assume that the direct labor costs includes fringe benefits, $25 per hour is not out of the question. This brings the direct labor cost per car to about $1000. per car.

2. Using the $800 estimate and dividing that by $20/hour yields 40 hours, a bit low but in the ball park.

3. $$\frac{\$10/hr \ \times \ 40 \ hr/wk \ \times \ 50 \ wk/yr \ \times \ 750}{24.5 \ \times \ 10^6} = \frac{15}{24.5} = 61\%$$

   This is 61% of factory costs which usually include only direct labor, direct material, and power as part of factory overhead. The total cost to run this business would be around $150 million.

4. In problem 4, the student must figure out what the power costs represents. The number in the table represents the annual cost for power based on the cost of $1 to $2 per square foot of space per year. Thus, a 300,000 sq. ft. building would cost around $600,000 per year based on the cost of $2 per square foot per year. The cost of power is usually only 1% to around 2% of the total cost to run a factory, assuming the factory doesn't use excessive power in the manufacturing processes (like a foundry with large electric furnaces).

CASE STUDY - CHAPTER 1

This is really an open-ended library-type assignment designed to expose students to the history of manufacturing and the various contributions that have significantly shaped and changed the discipline.

Chapter 2

PROPERTIES OF MATERIALS

1.   Metallic materials typically possess the properties of luster, high thermal conductivity, high electrical conductivity, and ductility.

2.   Some common physical properties of metals include: density or weight, melting point, optical properties (such as transparency, opaqueness or color), thermal properties (such as specific heat, coefficient of thermal expansion and thermal conductivity), electrical conductivity and magnetic properties.

3.   The results of standard tests apply only to the specific test conditions that were employed. Since actual service conditions rarely duplicate the conditions of laboratory testing, caution should be employed.

4.   The standard units for reporting stress in the English system is pounds per square inch (psi) , and in the metric system, it is megapascals (MPa). Being the ratio of one length to another length, strain is a dimensionless number. However, it is usually reported in terms of millimeter per meter, inch per inch, or strict percentage.

5.   Some of the important properties that relate to the elastic response of materials include: the proportionality limit - the stress at which the proportionality between stress and strain ceases to exist; Young's modulus or the modulus of elasticity which is the ratio of stress to strain in this region; the elastic limit - the maximum stress for which truly elastic behavior exists, and the resilience or modulus of resilience, which is a measure of the energy that a unit volume of material can absorb while in the elastic condition.

6.   The elastic-to-plastic transition can be designated in a variety of ways. If the transition is a distinct one, it is known as yield point, with the highest stress preceding the plastic strain being called the upper yield point, and the lower, "runout" value is the lower yield point. If the transition is not distinct, it is DEFINED through the concept of offset yield strength, the value of the stress associated with a specified, but tolerable, amount of plastic strain.

7.   The tensile test properties that relate directly to the plastic deformation behavior of materials include: the percent elongation (both uniform elongation and total elongation at fracture), percent reduction in area, and the strain-hardening coefficient.

8.   Ductility is generally evaluated by measuring the percent elongation (elongation at fracture) or the reduction in area at the narrowest point in the neck of the fractured specimen.

9.   In many cases, material "failure" is defined as the onset of localized deformation or necking. Since additional plastic deformation after necking would occur after "failure", it would be more appropriate to measure and report the uniform elongation (or the percent elongation prior to necking).

10.   Brittleness should not be equated with a lack of strength. Brittleness is simply the absence of significant plasticity. Many brittle materials, such as glass and ceramics can be used to impart significant strength to reinforced composites.

11.   Toughness is defined as the work per unit volume required to fracture a material, and can be used as one measure of a material's ability to absorb energy or impacts without cracking or breaking. Plastic deformation occurs during the measurement of toughness, whereas resilience requires the material to remain elastic.

12.   True stress considers the load as being supported by the actual area of the specimen and is a true indication of the internal pressures. Engineering stress is simply a normalizing of the load, dividing it by the original cross-sectional area of the specimen, i.e. dividing it by a constant. While easy to obtain, the engineering stress has little, if any, physical significance when the actual area is different from the original.
     The true, natural, or logarithmic strain is calculated by taking the natural logarithm of the current length divided by the original length, which is the sum of all of the incremental changes in length divided by the instantaneous length. It has the attractive property of being additive, i.e. the sum of the incremental strains is equal to the total strain from start to finish. Engineering strain, on the other hand, simply divides the elongation by a constant, the original length. While mathematically simple, the resultant value is not additive and has meaning only in reference to the original shape.

13.   Strain hardening or work hardening is the term used to describe the phenomenon that most metals actually become stronger and harder when plastically deformed. In deformation processes, this means that further deformation will generally require greater forces than those required for the initial deformation. Moreover, the product will emerge stronger than the starting material. From a manufacturing perspective, this means that the material is becoming stronger as it is being converted into a more useful shape -- a double benefit. One method of measuring and reporting this behavior is through the strain hardening exponent, n, which is obtained by fitting the true stress-true strain data to the equation form:
     $$\sigma = K \varepsilon^n$$

14.   The hardness of materials has often been associated with
the resistance to permanent indentation under the conditions of
static or dynamic loading. Other phenomena related to hardness
include the resistance to scratching, energy absorption under
impact loading, wear resistance, and resistance to cutting or
drilling.

15.   Some of the limitations of the Brinell hardness test
include: (1) it cannot be used on very hard or very soft
materials, (2) the results may not be valid for thin specimens,
(3) the test is not valid for case-hardened surfaces, (4) tests
near material edges may be invalid, (5) the large indentation may
be objectionable on finished parts, and (6) the edge of the
indentation may not always be well-defined or clearly visible.

16.   The Brinell hardness test is a two-step test that first
produces an indentation and then determines the <u>area</u> of the
permanent deformation mark. The computed Brinell hardness
has physical significance as the load per unit area in the
indentation. The Rockwell test correlates its results with the
<u>depth</u> of the indentation. Hardness numbers are assigned on the
basis of this depth, but the actual numbers have no absolute
significance, acquiring meaning only by basis of comparison.

17.   The various microhardness tests have been developed for
applications where it is necessary to determine the hardness of a
very small area of material or the hardness of thin material
where one wishes to avoid any interaction with the opposing
surface and support material.

18.   There are a wide variety of hardness tests and they often
evaluate different phenomenon: i.e. resistance to permanent or
plastic deformation, scratch or wear resistance, rebound energy,
and elastic deformation. All results are termed "hardness", but
little correlation is expected.

19.   There is often a direct correlation between penetration
hardness and tensile strength. For plain carbon and low-alloy
steels, the tensile strength in pounds per square inch can be
estimated by multiplying the Brinell hardness number by 500. For
other materials, the relationship may be different.

20.   The compression test is more difficult to conduct than the
standard tensile test. Test specimens must have larger cross-
sectional areas to resist buckling. As deformation proceeds, the
cross section of the specimen increases, producing a substantial
increase in the required load. Frictional effects between the
testing machine surfaces and the end surfaces of the specimen
will tend to alter the results if not properly considered.

21.   Static loads are ones that do not change over time, such as
the weight of a structure. Dynamic loads change over time, and
may take the form of: impacts, cyclic loads, or changes in mode
of loading (such as compression to tension).

22.   The two most common bending impact tests are the Charpy test and the Izod test. The Charpy test loads the specimen (usually notched) in three-point bending. The Izod test loads the specimen in a cantilever fashion.

23.   Designers should use extreme caution when applying impact test data for design purposes because the test results apply only to standard specimens containing a standard notch loaded under one condition of impact rate. Modifications in specimen size, the size and shape of the notch, and speed of the impact can produce significant changes in the results.

24.   There are several mechanisms by which metals can fracture when exposed to stresses that are below the yield strength. Cyclically-loaded materials can fail under repeated applications of stresses below the yield strength by a mechanism known as fatigue. Failure can also occur under conditions of elevated temperature and loads below the yield strength. The mechanism here is one of creep failure.

25.   Fatigue strength is the stress that a fatigue specimen was capable of withstanding for a specified number of load cycles, and therefore refers to any point on a standard S-N plot. Endurance limit or endurance strength, on the other hand, is the limiting stress level below which the material will not fail regardless of the number of cycles of loading.

26.   Several factors can drastically alter the fatigue properties of a material. One dominant factor is the presence of stress raisers, such as small surface cracks, machining marks, or gouges. Other factors include the temperature of testing, variation in the testing environment (such as humidity or corrosive atmosphere), residual stresses, and variations in the applied load during the service history.

27.   For steels, the endurance limit can be approximated as 0.5 times the ultimate tensile strength as determined by a standard tensile test.

28.   The initiation of a fatigue crack usually corresponds to discontinuities in the material. These may take the form of a fine surface crack, a sharp corner, machining marks, or "metal-lurgical notches", such as an abrupt change in metal structure.

29.   Engineered products frequently operate over a range of temperatures and often have to endure temperature extremes. The materials that are used in these products must exhibit the desired mechanical and physical properties over this range of temperatures. Thus, it is imperative that the designers consider both the short-range and long-range effects of temperature on the materials. This is particularly important when one realizes that the bulk of tabulated material data refers to properties and characteristics at room temperature.

30.   Steels and other body-centered crystal structure metals exhibit a ductile-to-brittle transition upon cooling. If this transition occurs at temperatures above those of service, the material will be used in a brittle condition, and sudden, unexpected fractures can occur under conditions that the material would be expected to endure.

31.   Material behavior under long-time exposure to elevated temperature is generally evaluated through creep testing, wherein a tensile specimen is subjected to fixed load at elevated temperature. Single tests provide data relating to the rate of elongation and the time to rupture under the specific conditions of testing. A composite of various tests can be used to evaluate the creep rate or rupture life under a variety of load and temperature conditions.

32.   The creep rate is the slope of the strain versus time curve in the long second stage region where the elongation rate is somewhat linear.

33.   Machinability, formability, and weldability all refer to the way in which a material responds to specific processing techniques. In most of these areas, however, there is a wide variation in the types and details of the processing. For example, forming includes bulk plastic deformation, localized bending, cold forming, hot forming, high speed deformation, and others. A favorable response to one condition of forming may not indicate a favorable response to others. Likewise, a material that is easily arc welded, may experience difficulties with the oxyacetylene methods.

34.   The basic premise of the fracture mechanics approach to testing and design is that all materials contain flaws or defects of some given size. Fracture mechanics then attempts to distinguish between the conditions where these defects will remain dormant and those conditions for which the defects might grow and propagate to failure.

35.   The three principal quantities that fracture mechanics tries to relate are: (1) the size of the largest or most critical flaw, (2) the applied stress, and (3) the fracture toughness of the material (a material property).

36.   The three primary thermal properties of a material are: (1) heat capacity or specific heat - a measure of the amount of energy that must be imparted or extracted to produce a one degree change in temperature; (2) thermal conductivity - a measure of the rate at which heat can be transported or conducted through a material; and (3) thermal expansion - a measure of the degree of expansion or contraction that will occur upon heating or cooling of the material.

CASE STUDY - CHAPTER 2
Separating Mixed Materials

a). Since both materials are in the hot-rolled condition, consultation with a materials handbook or other steel reference will indicate that the strength and hardness will differ for the different carbon contents (the 1020 steel contains 0.20% carbon, and the 1040 steel contains 0.40% carbon). A Rockwell hardness test is capable of distinguishing between these two materials and is simple, quick and inexpensive. On the Rockwell B scale, the hot-rolled 1020 material should have a reading of 75-78, while the 1040 material would be in the neighborhood of 94-95.

b). While both of the alloys are stainless steels, Type 430 is ferritic and is ferromagnetic. Type 316 is austenitic and is not ferromagnetic. Type 430 will be strongly attracted to a magnet, while type 316 will not. A small permanent magnet should be sufficient to distinguish between these two materials.

c). The separation of aluminum and magnesium becomes more difficult because of the similarity of properties. The difference in density would probably be noticeable only when handling like-size pieces. Therefore, separation would most likely require the identification of a simple and inexpensive means of detecting a significant difference. Eddy current non-destructive testing provides a signal that is dependent upon the electrical conductivity of the material being tested, and these materials have sufficiently different conductivities to provide a clear discrimination. An alternative method might involve some form of corrosion spot test, such as a drop of mild acid. The aluminum should be corrosion resistant, and the magnesium would react.

d). Density is often a usable discriminant for polymeric materials. The density of polyethylene is approximately 0.92 grams per cubic centimeter, and the density of polypropylene is approximately 0.98 grams per cubic centimeter. By creating a water-based solution with a density intermediate to these two values, the polymers can be separated by flotation -- the lighter polyethylene floating and the denser polypropylene sinking.

19.   Metallic crystals respond to low applied loads by simply stretching or compressing the distance between atoms. All atoms retain their basic positions, with the load serving only to disrupt the force balance of the atomic bonds in such a way as to produce elastic deformations.

20.   Plastic deformation is a permanent shift of atoms resulting in a permanent change in size or shape.

21.   A slip system for the plastic deformation of a metal is the specific combination of a preferred plane and a preferred direction within that plane. In general, the preferred planes are those with the highest atomic density and greatest parallel separation - the close-packed planes. The preferred directions are the close-packed directions.

22.   The dominant mechanical property of the bcc crystal structure metals is high strength. The fcc metals have high ductility. The hcp metals tend to be brittle.

23.   A dislocation is a line-type defect within a crystalline solid. Edge dislocations are the terminal edges of extra half-planes of atoms, and screw dislocations are the ends of partial "tears" through the crystal. Since the movement of dislocations provides the plasticity of a material, the force required to move dislocations determines the resistance to plastic deformation, or the strength of the material.

24.   Other crystal imperfections can provide effective barriers to dislocation movement and be used to strengthen the metal. These include: point-type defects (such as vacancies, interstitials, or substitutional atoms), additional line-type dislocations, and surface-type defects (such as grain boundaries).

25.   The three major types of point defects in crystalline materials are: vacancies (missing atoms), interstitials (extra atoms forced between regular atom sites), and substitutional atoms (atoms of a different variety occupying lattice sites).

26.   The strain hardening of a metal is the result of the multiplication of the number of dislocations and the interaction between the various dislocations to pin or block the movements of one another.

27.   Since dislocations cannot cross grain boundaries (a discontinuity to crystal structure), these boundaries serve to impede dislocation movement and make the material stronger. A material with a finer grain structure (more grain boundaries) will, therefore, tend to be stronger than one with larger grains.

28.   When metals are deformed, the crystals cease to be spherically symmetric, and become elongated in the direction of metal flow. One result may be the creation of anisotropic properties --properties that vary with direction.

29.   Brittle fractures occur without the prior warning of plastic deformation and propagate rapidly through the metal with little energy absorption. Ductile fractures generally occur after the available plastic deformation has been exceeded.

30.   Plastic deformation increases the internal energy of a material through both the creation of numerous additional dislocations and the increased surface area of the distorted grain boundaries. Given the opportunity, the metal will seek to reduce its energy through the creation of a new crystal structure, i.e. recrystallize.

31.   Recrystallization is often used to restore ductility to a metal and enable further deformation to be performed. Without recrystallization, further deformation would result in fracture. NOTE: If the deformation is performed at temperatures above the recrystallization temperature, deformation and recrystallization can take place simultaneously and large deformations are possible.

32.   The major distinguishing factor between hot and cold working is whether the deformation is produced at a temperature that is above or below the recrystallization temperature of the metal. In cold working, no recrystallization occurs and the metal retains its strain hardened condition. When hot working is performed, recrystallization produces a new grain structure and no strain hardening is possible.

33.   When an alloy addition is made to a base metal, several possibilities can occur. The two materials can be insoluble and refuse to combine or interact. If there is solubility, the alloy can dissolve in the base metal to produce a solid solution of either the substitutional or interstitial variety. A final possibility is that the two can react to produce an intermetallic compound - a combination with definite atomic proportions and definite geometric relationships.

34.   Intermetallic compounds tend to be hard, brittle, high-strength materials.

35.   Electrical resistance in a metal depends largely on two factors - the number of lattice imperfections and the temperature. Vacancies, interstitials, substitutional atoms, dislocations, and grain boundaries all act as disruptions to the regularity of a crystalline lattice. Thermal energy causes the atoms to vibrate about their equilibrium positions and interferes with electron transport.

36.   Intrinsic semiconductors are ones that occur naturally. Extrinsic semiconductors have chemistries that have been modified by "doping" to enhance or alter their conductivity.

Wood (such as kiln-dried Ponderosa pine) is easily shaped, can be painted or finished in a wide spectrum of finishes, and has low thermal conductivity (keeping the winter cold and summer heat out). Unfortunately, the material has a definite grain structure, which may lead to cracking or splintering. The material requires special impregnation and coating to improve its ability to resist degradation. Wood requires regular surface maintenance (such as painting or sealing) to minimize moisture absorption and rot. While its dimensions are relatively insensitive to changes in temperature, they can change significantly with changes in humidity or moisture content, leading to possible warping or twisting. The shrinking, swelling and cracking tendencies make it extremely difficult to provide a durable surface protection. Finally, wood is a combustible material.

Aluminum can be extruded into the complex channels used for window frames, is durable, non-corrosive, and can be color anodized or finished into a variety of surfaces. The properties are consistent and predictable and do not change over time, or with variations in temperature (over the range where windows would operate). The material does not absorb moisture, swell, shrink, split, crack or rust. Maintenance is extremely low, but the material has a high thermal conductivity. If the same piece is exposed to a cold exterior and warm, moist interior (as in winter weather), the material will try to achieve thermal uniformity. The inside surfaces will "sweat" with condensation, and thermal efficiency of the window will be poor. Compared with alternatives, however, aluminum is stronger and more rigid (23 times stiffer than vinyl). From a safety perspective, aluminum is noncombustible and does not emit any toxic fumes when heated to high temperature.

Vinyl windows offer a range of color, and the color is integral to the material. There is no need for any surface finishing and the appearance requires no periodic maintenance. In addition, the thermal conductivity is low, giving the window good thermal efficiency. Unfortunately, polymers have poor dimensional stability, generally shrinking over time, and often deteriorate with prolonged exposure to ultraviolet light (becoming brittle). Since windows will see prolonged exposure to sunlight, the long term durability and stability may come into question. The thermal expansion of vinyl is considerably greater than either aluminum or wood, and the resulting dimensional changes may cause distortion of the windows. In addition, the properties of vinyl will vary over the temperature range that the product will see. When heated, vinyl loses strength, and when cold, it becomes more brittle and less impact resistant. The material is combustible and may emit toxic fumes when exposed to high temperatures.

It would appear that aluminum is a superior structural material, whose primary detriment is its high thermal conductivity. If a design could be developed to insert some form of conductivity barrier between the outside and inside surfaces, the resulting window would offer the best of all worlds. Several companies currently offer such a design, linking the inside and outside extrusions with a high-strength polymeric link. Being totally internal, this polymer is not subject to sunlight deterioration, and does not significantly impair the structural performance of the window.

Chapter 4

EQUILIBRIUM DIAGRAMS AND THE IRON-CARBON SYSTEM

1.   A phase is a portion of a substance possessing a well-defined structure, uniform composition, and distinct boundaries or interfaces.

2.   In a glass of soda with ice, the soda is continuous and the ice is discontinuous. Helium in a balloon is a gaseous phase, and coffee with cream is a single-phase solution.

3.   An equilibrium phase diagram is a graphical mapping of the natural tendencies of a material system (assuming that equilibrium has been attained) as a function of such variables as pressure, temperature, and composition.

4.   The three primary variables considered in equilibrium phase diagrams are: temperature, pressure and composition.

5.   A pressure-temperature phase diagram is not that useful for many engineering applications because most processes are conducted at atmospheric pressure. Most variations occur in temperature and composition.

6.   A cooling curve is a temperature versus time plot of the cooling history when a fixed-composition material is heated and subsequently cooled by removing heat at a uniformly slow rate.

7.   Transitions in a material's structure are indicated by characteristic points on the cooling curve. These characteristic points may take the form of an isothermal hold, abrupt change in slope, or localized aberration to the continuity of the curve.

8.   Solubility limits denote the conditions at which a solution becomes completely saturated, i.e. any additional solute must go into a second phase. Solubility limits are generally determined through use of inspection techniques such as X-ray analysis (detects where a new crystal structure or lattice spacing appears) or microscopy (detects the presence of the second phase), that can be used to identify the composition where the transition from one to two-phase occurs.

9.   In general, as the temperature of a system is increased, the maximum amount of a substance that can be held in solution also increases.

10. Upon crossing the liquidus line during cooling, the first solid begins to form in the material. Upon crossing the solidus line, solidification is complete, i.e. there is no longer any liquid present. Upon crossing a solvus line, a single phase material begins to precipitate a second phase, since the solubility limit is now being exceeded.

11.   The three pieces of information that can be obtained from each point in an equilibrium phase diagram are: the phases present, the composition (or chemistry) of each phase, and the amount of each phase present.

12.   A tie-line is an isothermal line drawn through any point in the two phase region of a phase diagram, terminating at the boundaries of the single phase regions on either side. It is used in the two-phase regions of an equilibrium phase diagram.

13.   The end points of the tie-line correspond to the compositions of the two phases present.

14.   The relative amounts of the component phases in a two-phase mixture can be computed through use of the lever law. The tie-line is separated into two segments by dividing it at the chemistry of the alloy in question. The fraction of the total length of the tie-line that lies opposite to a given phase corresponds to the fractional amount of that particular phase.

15.   As a metal solidifies through a freezing range, the chemistry of the solid phase is constantly changing. With rapid cooling, the material diffusion within the material is insufficient to achieve uniform chemistry. The various regions of the solid possess the chemistry characteristic of the temperature at which they solidified and a "cored" structure results.

16.   Three-phase reactions appear as horizontal lines in binary (two-component) phase diagrams. These lines have a distinctive V intersecting from above, or an inverted V intersecting from below. The intersection of the V and the horizontal line denotes the three-phase reaction, which is usually written in the form of cooling, i.e. the phases present above the line going to those present below.

17.   A eutectic reaction has the general form of Liquid --->
Solid 1 + Solid 2. In essence, a liquid solidifies to form two distinctly different solids of differing chemistries.

18.   Eutectic alloys are attractive for casting and as filler metals in soldering and brazing because they generally have the lowest melting point of all alloys in a given system and solidify into a relatively high-strength structure.

19.   A stoichiometric intermetallic compound is a single-phase solid that forms when two elements react to form a compound of fixed atomic ratio. The compound cannot tolerate any deviation from that fixed ratio, so it appears as a single vertical line in a phase diagram, breaking the diagram into recognizable subareas. Non-stoichiometric intermetallic compounds are single phases that appear in the central regions of a phase diagram, that can tolerate chemical variations, and thus have an observed width. They appear as a region and not a line.

20.   In general, intermetallic compounds tend to be hard, brittle materials.

21.   If an intermetallic compound can be uniformly distributed throughout a structure in the form of small particles in a ductile matrix, the effect can be considerable strengthening of the material. If the intermetallic should become the continuous phase (as in a grain boundary coating) or be present in large quantities, the material will be characteristically brittle.

22.   The four single phases in the iron-carbon equilibrium phase diagram are: ferrite (alpha), which is the room-temperature body-centered cubic structure; austenite (gamma), the elevated temperature face-centered cubic phase; delta-ferrite (delta), the high-temperature body-centered cubic phase; and cementite (Fe_3C), the iron-carbon intermetallic compound that occurs at 6.67 wt. percent carbon.

23.   The point of maximum carbon solubility in iron, 2.11 weight percent, forms an arbitrary division between steels and cast irons. Cast irons contain greater than 2.11% carbon and experience a eutectic reaction upon cooling.

24.   Some of the key characteristics of austenite are its high formability (characteristic of the fcc crystal structure) and its high solubility of carbon (a good starting point for heat treatment).

25.   The most important three-phase reaction in the iron-carbon diagram when considering steels is certainly the eutectoid reaction. Under equilibrium conditions, austenite of 0.77 weight percent carbon and the fcc crystal structure transforms into ferrite of the bcc crystal structure, capable of holding only 0.02% carbon and cementite or iron carbide with 6.67% carbon. In essence, iron changes crystal structure and the rejected carbon goes to form the iron carbide intermetallic.

26.   The fcc crystal structure of austenite is capable of dissolving as much as 2.11% carbon at elevated temperature. In contrast, the bcc crystal structure of ferrite can hold only 0.02% carbon at its maximum solubility and 0.007% at room temperature.

27.   Pearlite is the name given to the structure formed when austenite undergoes the eutectoid reaction under equilibrium (or near-equilibrium) conditions. It is a lamellar structure composed of alternating plates of ferrite and cementite, but has its own characteristic set of properties, since it always forms from the same chemistry at the same temperature.

28.   Steels having less than the eutectoid amount of carbon (less than 0.77% carbon) are called hypoeutectoid steels. Their structure consists of regions of ferrite that formed before the eutectoid reaction (primary or proeutectoid ferrite) and pearlite that formed as the remaining austenite underwent the eutectoid transformation. Steels with greater than 0.77% carbon are called hypereutectoid steels and have structures consisting of primary cementite and pearlite.

29.   The general composition of cast irons is 2.0 to 4.0% carbon, 0.5 to 3.0% silicon, less than 1.0% manganese, and less than 0.2% sulfur. In addition, nickel, copper, chromium and molybdenum may be added as alloys. Silicon is the major new addition. It partially substitutes for carbon, and promotes the formation of graphite as the high-carbon phase.

30.   Cast irons often contain graphite as the high-carbon phase instead of the cementite (or iron carbide) commonly found in steels. Graphite formation is promoted by slow cooling, high carbon and silicon contents, heavy section sizes, inoculation practices, and alloy additions of Ni and Cu. Cementite is favored by fast cooling, low carbon and silicon levels, thin sections, and alloy additions of Mn, Cr, and Mo.

31.   The microstructure of gray cast iron consists of three-dimensional graphite flakes dispersed in a matrix of ferrite, pearlite, or other iron-based structure.

32.   Since the graphite flakes in gray cast iron have no appreciable strength, efforts to increase the strength of this material must focus on improving the strength of the iron-based matrix structure.

33.   Gray cast irons possess excellent compressive strengths, excellent machinability, good wear resistance, and outstanding vibration damping characteristics. In addition, the silicon provides good corrosion resistance and the high fluidity desired for castings. Low cost is an additional asset.

34.   White cast iron is very hard, but very brittle. It finds application where extreme wear resistance is required.

35.   Malleable iron is essentially heat-treated white cast iron where a long time thermal treatment changes the carbon-rich phase from cementite to irregular graphite spheroids. The more favorable graphite shape removes the internal notches of gray cast iron and imparts the increased ductility and fracture resistance.

36.   In ductile cast iron, the cost of the nodulizer, higher-grade melting stock, better furnaces, and improved process control required for its manufacture all contribute to an increased cost over materials such as gray iron.

37.   Compacted graphite cast iron is characterized by a graphite structure intermediate to the flake graphite of gray cast iron and the nodular graphite of ductile iron. It forms directly upon solidification, and possesses some of the desirable properties and characteristics of each.

CASE STUDY - CHAPTER 4
The Blacksmith Anvils

1.   The anvil will be subjected to the shock of direct and indirect hammer blows during the forging of metals. The surfaces must be resistant to wear, deformation and chipping, and good energy absorption or damping characteristics would be an added plus, acting to reduce noise and vibration. The anvil must have sufficient mass to absorb the blows, and not tip or move when the blows are offset from the base. The material must resist damage when red-hot metal is placed in contact with its surfaces for brief to moderate periods of time. Since heat retention in the workpiece is desirable, heat transfer to and through the anvil should be minimized. Corrosion resistance to a normal shop atmosphere would also be desirable. While the dimensional requirements can be somewhat lenient, the working surface should be flat and reasonably smooth.

2.   Features influencing the method of fabrication include the cited yield strength, elongation and hardness, and somewhat limited production quantity. In addition, the somewhat massive size (both weight and thickness) can be quite restrictive. The width is probably about 4-inches or greater. Other than a single mirror plane, there is no significant symmetry or uniformity of cross section. Handling should be minimized because of the size and weight.

Because of the size (both length and thickness), complexity of shape, and limited production quantity, some form of expendable-mold casting appears to be the most attractive process. Forging would be another alternative, but the size and quantity could be quite limiting.

3.   Because of the need for impact resistance, cast irons would most likely come from either the malleable or ductile families, but the section thickness may present problems for the production of malleable. Cast steels would also be quite attractive, but the higher melting temperatures, lower fluidity, and high shrinkage could present problems. Alloyed material may be necessary for the desired heat treatment response.

4.   Production alternatives include casting the entire piece from a single material, or casting the base of one material (including the horn) and welding or otherwise attaching a plate of stronger material to the top. Because of the desire to replicate the 1870's design, a single material is probably preferred.

While any of the materials discussed in section 3 would be workable alternatives, ductile cast iron might be the most attractive. It would most likely be cast in some form of sand mold, possibly one with higher strength than green sand. After casting, heat treatment would likely be necessary to establish the desired properties. A normalizing or annealing treatment would produce the desired properties with a very stable structure. The working surfaces and mounting base would be subjected to some form of surface grinding. If needed, a deep surface hardening treatment, such as flame hardening, could be used on the critical surfaces to increase the hardness.

Chapter 5

HEAT TREATMENT

1.   Heat treatment is the controlled heating and cooling of metals for the purpose of altering their properties. Its importance as a manufacturing process stems from the extent to which properties can be altered.

2.   While the term "heat treatment" applies only to processes where the heating and cooling are done for the specific purpose of altering properties, heating and cooling often occur as incidental phases of other manufacturing processes, such as hot forming and welding. Material properties will be altered as the material responds in the same way it would if an intentional heat treatment had been performed. Properties can be significantly altered by the heating and cooling.

3.   Processing heat treatments are slow cool, rather long time, treatments designed to prepare a material for fabrication. Some possible goals of these treatments are: improve machining characteristics, reduce forming forces, or restore ductility for further fabrication.

4.   Since most processing heat treatments involve rather slow cooling or extended time at elevated temperature, the conditions tend to approximate equilibrium, and equilibrium phase diagrams can be used as a tool to understand and determine process details.

5.   Annealing operations may be performed for a number of reasons, among them: to reduce strength or hardness, remove residual stresses, improve toughness, restore ductility, refine grain size, reduce segregation, or alter the electrical or magnetic properties of a material.

6.   Full anneals can produce extremely soft and ductile structures, but they are time consuming and require considerable energy to maintain the elevated temperatures required during the soaking and furnace cooling. In addition, the furnace temperature is changed during the treatment, so the furnace must be reheated to start another cycle.

7.   If hypereutectoid steels were slow cooled from the all-austenite region, they would spend considerable time in the austenite plus cementite condition, and the hard, brittle cementite that forms would tend to produce a continuous network along the grain boundaries. A small amount of cementite in a continuous network can make the entire material brittle.

8.   The major difference of normalizing compared to full annealing is the use of an air cool in place of the long time, controlled furnace cool. This reduces processing time, furnace time, and fuel and energy use. However, the furnace cool of a full anneal imposes identical cooling conditions at all locations within the metal and produces identical properties. With normalizing, the cooling will be different at various locations. Properties will vary between surface and interior, and different thickness regions will have different properties.

9.   Process heat treatments that do not require the reausteni-tization of the steel include: the process anneal, designed to promote recrystallization and restore ductility; the stress-relief anneal, designed to remove residual stresses; and spheroidization, a process to produce a structure that enhances the machinability or formability of high-carbon steels.

10.   Process anneals are performed on low-carbon steels with carbon contents below 0.25% carbon. Spheroidization is employed on high-carbon steels with carbon contents greater than 0.6%C.

11.   The recrystallization process and its kinetics is a function of the particular metal, the degree of prior straining, and the time provided for completion. In general, the more a metal has been strained the more energy has been stored, and the lower the recrystallization temperature or the shorter the time.

12.   The six major mechanisms available to increase the strength of a metal are: solid solution strengthening, strain hardening, grain size refinement, precipitation hardening, dispersion hardening, and phase transformation hardening. All techniques are not applicable to every metal.

13.   The most effective strengthening mechanism for the nonferrous metals is precipitation hardening.

14.   Precipitation hardening begins with a solution treatment to create an elevated-temperature single-phase solid solution, followed by a rapid quench to produce a supersaturated solid solution, and then a controlled reheat to age the material (cause the material to move toward the formation of the stable two-phase structure).

15.   Precipitation hardening metals are either naturally aging (ages at room temperature) or artificially aging (requires elevated temperature to produce aging). Considerable flexibility and control is offered by artificial aging, since the properties can be altered and controlled by controlling the time and temperature of elevated temperature aging. Dropping the temperature terminates diffusion and retains the structure and properties present at that time. NOTE: Subsequent heating, however, will continue the aging process!

16.   In a coherent precipitate, the crystallographic planes of the parent structure are continuous through the precipitate cluster, and the solute aggregate tends to distort the lattice to a substantial surrounding region. In contrast, second-phase particles have their own crystal structure and distinct interphase boundaries.

17.   In constructing the IT or T-T-T diagram, thin specimens of a metal are heated to form uniform, single-phase austenite, and are then instantly quenched to a temperature where austenite is not the stable phase. The samples are then held at this constant temperature for variable periods of time and the kinetics of the structure change are determined. Such instantaneous changes in temperature followed by isothermal holds are quite unrealistic for manufactured items, which usually undergo some form of continuous cooling as heat is extracted from surfaces and fed to the surfaces from the hot interior.

18.   For steels below the A_1 temperature, the stable phases predicted by the equilibrium phase diagram are ferrite and cementite.

19.   According to the T-T-T diagram, some of the non-equilibrium structures that may be present in heat-treated steels are: bainite, martensite, tempered martensite, and retained austenite.

20.   Martensite forms from austenite by an instantaneous change in crystal structure with no diffusion. The fcc austenite transforms to the body-centered structure which is distorted into a tetragonal shape to accommodate the additional carbon. The degree of distortion is proportional to the amount of trapped carbon.

21.   The major factor determining the strength and hardness of steel in the martensitic structure is the amount of carbon present in the steel.

22.   Retained austenite is austenite which remains in a metastable state at temperatures where the equilibrium phase diagram predicts that it should no longer exist. It can be responsible for low strength or hardness, dimensional instability or cracking, or brittleness (by transforming to untempered martensite at some later time).

23.   As formed, martensite lacks sufficient toughness and ductility to be useful as an engineering material. Tempering is the controlled decomposition of the single-phase supersaturated solid solution toward the formation of the stable ferrite and cementite structure. Ductility and toughness improve at the expense of strength and hardness.

24.   Both heat treatments begin by replacing the original structure with an elevated-temperature, single-phase solid solution (redissolving any second phases). A quench then produces a supersaturated solid solution. A more moderate reheating then permits diffusion to move the material toward formation of the stable two-phase configuration. When age hardening, the quenched material is weaker and more ductile, and aging increases the strength at the expense of ductility. With the quench-and-temper process, the quenched structure is strong, but lacks ductility. Tempering increases ductility at the expense of strength.

25.   T-T-T diagrams are usually not applicable to direct engineering utilization because the assumptions of instantaneous cooling followed by complete isothermal transformation fail to match reality. The C-C-T diagram maps the changes that occur as a metal is continuously cooled at various rates of cooling.

26.   In the Jominy end-quench hardenability test, a standard steel specimen is subjected to a standardized quench. Since the thermal conductivity of steel is essentially constant for the range of carbon and low-alloy steels, the cooling rate varies with the distance from the quenched end - from rapid quench to an approximate air-cool.

27.   The quench in the Jominy test is standardized by specifying the quench medium (water), quenchant temperature (75 F), internal nozzle diameter (1/2 inch), water pressure, and the gap between the nozzle and the specimen.

28.   The concept of "equivalent cooling rates" is based on the assumption that identical results will be obtained if a material undergoes identical cooling history. If the cooling rate is known for a given location within a part, the properties at that location can be predicted as those at the equivalent cooling rate location of a Jominy test bar (well-documented in many reference texts).

29.   Hardenability is a measure of the depth to which full hardness can be obtained when heat-treating a steel. It is primarily dependent upon the types and amounts of alloy elements in the steel.

30.   The depth of hardening can be increased by increasing either the severity of the quench or the hardenability of the steel. Quench changes may be limited by cracking or warping problems, however. Hardenability is increased by increasing the amount of alloy additions.

31.   The three stages of liquid quenching are: the vapor-jacket stage, the second stage in which the quenchant extracts heat by boiling, and the third stage where the mechanism of heat transfer is limited to conduction and convection.

32.   As a quench medium, water offers a high heat of vaporization, and second-stage cooling down to 212°F. It is cheap,
readily available, easily stored, nontoxic, nonflammable,
smokeless, and easy to filter and pump. On the negative side,
the bubbles tend to cling, it is an oxidizing medium, and is
corrosive. In addition, the rapid rates of cooling often induce
distortion and cracking.

33.   Oil quenches are generally less likely to produce quench
cracks than water or brine for several reasons. The rate of heat
extraction into boiling oil is slower than in boiling water. The
major difference, however, is due to the fact that the boiling
points of oils are sufficiently high that the transition to the
third-stage of quenching occurs before the martensite start
temperature. Slower cooling through the martensite transformation leads to a milder temperature gradient and a reduced
likelihood of cracking.

34.   Polymer or synthetic quenches cool more rapidly than oils
but slower than water or brine. They can be tailored by varying
the concentrations of the quench components to provide extremely
uniform and reproducible results. They are less corrosive than
water or brine, are cheaper and less of a fire hazard compared to
oils, and tend to minimize distortion.

35.   Some undesirable design features in parts that are to be
heat treated include: nonuniform sections or thicknesses, sharp
interior corners, and sharp exterior corners.

36.   When steel is quenched, the elevated temperature
face-centered cubic structure changes to the body-centered
configuration, and EXPANDS! When aluminum is quenched, it cools
and thermally CONTRACTS. In most cases, the residual stresses
formed by cooling tend to be opposite for the two materials.

37.   Residual stresses can be undesirable because, in service,
they add algebraically to the stresses applied to the part. Loads
well within the design limit may couple with unfavorable residual
stresses to produce failure. By themselves, residual stresses
may produce unwanted distortion or cracking.

38.   Two methods of producing strong structures while minimizing
residual stresses and the likelihood of cracking are austempering
and martempering. A rapid cool is used to reduce the temperature
of the material to just above the martensite start. The
temperature is then allowed to become uniform prior to either
further cooling to martensite (martemper) or isothermal
transformation to bainite (austemper).

39.   In thermomechanical processing, mechanical deformation and
heat treatment are intimately combined into a single process.

40.   Selective heating techniques for surface hardening include: flame hardening, induction hardening, laser beam hardening, electron beam hardening, and lead-pot or salt bath immersion.

41.   Laser beam surface hardening operates at high speeds, produces little distortion, induces compressive residual stresses on the surface, and can be used to harden selected surface areas while leaving the remaining surfaces unaffected. Computer control and automation can be readily used and conventional mirrors and optics can be used to shape and manipulate the beam.

42.   In pack carburizing, the heating is inefficient, temperature uniformity is questionable, handling is often difficult, and the process is not readily adaptable to continuous operation. By replacing the solid carburizing compound with a carbon-providing gas, the operation becomes faster and more easily controlled. Accuracy and uniformity are increased, and continuous operation is possible.

43.   Compared to conventional nitriding or carburizing, ionitriding offers shorter cycle times, reduced consumption of gases, significantly reduced energy costs, reduced space requirements and the possibility of total automation. Product quality is improved and the process is applicable to a wider range of materials.

44.   Batch furnaces may be preferred to continuous furnaces when the productions runs are small and the details of the thermal processing vary from lot to lot. Continuous furnaces are best for large production runs of the same or similar parts that undergo the same thermal process.

45.   Artificial atmospheres are often used during heat treatment operations to suppress undesirable reactions such as scale formation or tarnishing, prevent decarburization, or supply carbon or nitrogen for surface modification.

46.   In a fluidized-bed furnace, a bed of dry, inert particles is heated and fluidized by a stream of flowing gas. Parts introduced into the fluidized media become engulfed and are heated by radiant heating. Temperature and atmosphere can be altered quickly, heat transfer rate and thermal efficiency are high, and fuel consumption is low. Due to high flexibility, one furnace can be used for multiple applications.

47.   While heat treatment consumes large amounts of energy, its use may actually be an energy conservation measure because it enables the manufacture of a higher-quality, more durable, product. In addition, higher strengths may permit the use of less material to produce a comparable product.

PROBLEMS FOR CHAPTER 5

1.   While this is essentially a library-research project, it is
hoped that the student will note such features as the following:
Flame and induction hardening are performed on materials that
have the capability of possessing both the desired substrate
properties and the desired surface properties. Carburizing
alters the surface chemistry and achieves the desired hardness
through subsequent heat treatment. This treatment can take the
form of a direct quench from the carburizing treatment, a quench
from a reheat to a lower temperature, or dual surface and
substrate treatments. Nitriding also modifies the surface
chemistry, but the nitrided layer cannot sustain subsequent heat
treatment. Therefore, the substrate is fully heat-treated prior
to nitriding, and the hard surface is formed after the heat
treatment.
       The additional information can be found in numerous
references, such as Metals Handbook, or the references cited
under the "Heat Treatment" and "Surfaces and Finishes" sections
of the Selected References for Additional Study.

2.   While the basic information on these processes has been
summarized in the text, this problem encourages the students to
dig deeper in a library-research mode. One will learn, for
example, that there are actually several means of spheroidizing a
high-carbon steel. These are different processes with the same
objective and utilize the same name. It is important that users
understand the entire process, and all of the intricacies, such
as the different effects of full anneal and normalizing on
subsequent machining (as discussed in the text).
       Useful references again are Metals Handbook and those
listed in the "Heat Treatment" section of the Selected References
for Additional Study.

CASE STUDY - CHAPTER 5
A Flying Chip from a Sledgehammer

       The hammerhead chipped because of the formation of
untempered martensite. Untempered martensite of 0.6% carbon
would have a hardness of about Rockwell C 65 and would be
extremely hard and brittle, quite likely to crack upon impact.

       The procedure used to grind off the mushroom would likely
involve removing the handle to permit hand grinding of the head.
This would then be periodically dipped into a container of water
when it gets too hot to hold. Considering the size and mass of a
15-pound sledge hammer head, and the thermal conductivity of
steel, it is quite possible that the temperatures in the grinding
region could be sufficient to reaustenitize (>1333°F) the metal
before the operator would feel uncomfortably high temperatures in
the gripped region (especially if he were wearing some type of
protective leather-palmed glove). Upon water quench, the
austenite would transform to untempered martensite. Subsequent

grinding might bring about some degree of tempering, but this is not assured and all of the untempered martensite may not be affected.

Possible solutions to the problem include: (1) alteration of the grinding procedure to prevent the generation of such excessive temperatures, or (2) a required furnace retempering of the entire head prior to reassembly and reuse, and (3) mandatory use of safety goggles when using the sledge hammers.

Chapter 6

FERROUS METALS AND ALLOYS

1.   Many of the properties and characteristics of engineering materials depend not only on the material itself, but also on the manner of production and the details of processing. Aspects of prior processing can significantly influence both further processing and the final properties of the product.

2.   A ferrous material is one that is based on the element iron (i.e. iron is the major chemical constituent of the material).

3.   When iron ore is reduced to metallic iron, other elements are usually present in the product. All of the phosphorus and most of the manganese in the ore will also reduce and will enter the iron. The oxides of silicon and sulfur will be partially reduced and these elements will also become part of the metal.

4.   Pig iron is a high-carbon, high-silicon material with a chemistry in the range of 3.0 to 4.5% carbon, 1.0 to 3.0% silicon, 0.15 to 2.5% manganese, 0.05 to 0.1% sulfur, and 0.1 to 2.0% phosphorus. In the conversion into steel, the pig iron is subjected to an oxidation process that substantially decreases the amount of carbon, silicon, manganese, phosphorus, and sulfur.

5.   Ladle metallurgy refers to a variety of processes designed to provide final purification and fine tune both the chemistry and temperature of the melt. Alloy additions can be made, carbon can be further reduced, dissolved gases can be reduced or removed, and steps can be taken to control subsequent grain size, limit inclusion content, reduce sulfur, and control the shape of any included sulfides. Stirring, degassing, reheating, and various injection procedures can be performed to increase the cleanliness of the steel and provide tighter control of the chemistry and properties.

6.   By extracting molten steel from the bottom of a ladle, slag and floating matter are not transferred to the solidified product.

7.   Solidification shrinkage is the term applied to the often substantial change in dimensions that occurs when a liquid changes to solid. The more efficient arrangement of atoms results in an increase in density and a decrease in volume.

8.   While the amount of shrinkage cannot be changed, the shape and location can be greatly controlled. The amount of pipe can be reduced through the use of a ceramic hot top or exothermic topping. By retaining heat at the top, the liquid reservoir at the end of solidification is a more uniform layer, thereby minimizing the depth of the shrinkage cavity. Mold design, chills, and heat sources can all be used to alter the location of the last liquid to be present in a solidification process.

9.   Continuous casting virtually eliminates the problems of piping and mold spatter. In addition, it eliminates the pouring into molds, stripping the molds from the solidified metal, and the handling and reheating of the ingots prior to rolling. Cost, energy and scrap are all significantly reduced. The products have improved surfaces, more uniform chemical composition, and fewer oxide inclusions.

10.   Various alternatives exist to overcome the problems of dissolved gases in steel. Aluminum, ferromanganese or ferrosilicon can be added to the molten material to react with the dissolved oxygen and convert it to solid metallic oxides - a process known as deoxidation. Alternative approaches which actually remove most all of the various dissolved gases are vacuum degassing, vacuum arc remelting, vacuum induction melting, and electroslag remelting.

11.   Electroslag remelting can be used to produce extremely clean, gas-free metal. The nonmetallic impurities are collected in the flux blanket, leaving beneath a newly solidified structure with improved quality.

12.   A plain carbon steel is an alloy of iron and carbon, containing manganese, phosphorus, sulfur, and silicon in normal, but small, quantities.

13.   Medium-carbon steels are used in high volumes because they offer the best overall balance of engineering properties. The high fatigue and toughness properties of the low carbon steels are effectively compromised with the strength and hardness of the higher carbon contents.

14.   Plain-carbon steels are the lowest-cost steel material. Because of the low cost, they should be given first consideration for many applications.

15.   The most common alloy elements added to steel include: chromium, nickel, molybdenum, vanadium, tungsten, cobalt, boron, and copper, as well as manganese, phosphorus, sulfur, and silicon in amounts greater than normally present.

16.   Alloy elements are added to steel for a variety of reasons, among them: to improve the strength and hardenability, or to produce special properties, such as corrosion resistance or stability at high or low temperatures.

17.   Alloy elements that are particularly effective in increasing the hardenability of steel in order of decreasing effectiveness are: manganese, molybdenum, chromium, silicon, and nickel. Vanadium and boron are also used in small, but effective quantities.

18.   While many people feel that chromium is added to steel to improve the corrosion resistance, the primary purpose of a chromium addition is generally to increase hardenability and increase strength. Combined with carbon, chromium can also be used to improve wear resistance.

19.   Chromium, vanadium, molybdenum, and tungsten can all be used to impart strength and wear resistance through the formation of stable second-phase carbides.

20.   The last two digits in the AISI-SAE designation system for steel indicates the approximate carbon content of the steel in hundredths of weight percent. This is useful information since many engineering properties are directly tied to the carbon content.

21.   In selecting a steel, it is important to keep use and fabrication in mind. For example, a product that is to be assembled by welding would benefit from a lower carbon content as such would reduce the likelihood of cracking. Additional strength, if desired, would be better obtained through selection of additional alloy elements rather than an increase in the carbon content of the steel.

22.   There is a fundamental difference in the way strength is obtained in the HSLA and constructional alloy steels. The high strength/low alloy (HSLA) types rely largely on the chemical composition to develop the desired mechanical properties in the as-rolled or normalized condition. In contrast, the constructional alloys generally develop the desired properties through the use of nonequilibrium heat treatment.

23.   Microalloyed steels are steels that contain small amounts of alloying elements like niobium, vanadium, titanium, zirconium, boron, rare earth elements, or combinations thereof and are used as substitutes for heat-treated steels. Attractive strength and hardness is obtained without interfering with the material processing (weldability, machinability and formability).

24.   Microalloyed steels require less cold work to attain a desired level of strength, so the remaining ductility can be greater than with alternative materials. Hot formed products can often be used in the air cooled condition to provide properties comparable to quenched-and-tempered alloys. Machinability, fatigue life, and wear resistance can be superior to alternative materials. Energy savings can be substantial, straightening or stress relieving after heat treatment can be eliminated, and quench cracking is not a problem. Weight can often be reduced in parts, since the strength is increased.

25.   Free machining steels are basically carbon steels that have been modified by an alloy addition to enhance machinability. Sulfur, lead, bismuth, selenium, tellurium, and phosphorus have all been added to enhance machinability.

26.   When free-machining steels are selected, the ductility and impact properties are somewhat lower than with the unmodified steels.

27.   Bake-hardenable steels are aging resistant during normal storage, but begin to age during forming, and continue to age while exposed to heat during the paint baking operation. Since strengthening occurs after forming, the forming characteristics are good, coupled with improved product properties.

28.   Precoated steel sheets can often be used to offset the high cost of finishing products on a piece-by-piece basis -- a costly and time-consuming approach. Caution must be exercised to protect the coating during fabrication, but this is usually far less than the cost for finishing the individual pieces.

29.   The amorphous metals have attracted considerable attention for use in magnetic applications. Since the material has no grains or grain boundaries, the magnetic domains can move freely in response to magnetic fields, the properties are the same in all directions, and corrosion resistance is improved. The high magnetic strength and low hysteresis losses offer the possibility of smaller, lighter weight magnets.

30.   Maraging steels are used when super-high strength is the dominant requirement, and acceptable toughness is also needed. Yield strengths are often in excess of 250 ksi with elongations in excess of 11%.

31.   The corrosion resistance of stainless steels is the result of a strongly adherent chromium oxide that forms on the surface when the amount of chromium dissolved in the metal exceeds 12%.

32.   The ferritic stainless steels are the cheapest of the various families. If their properties are adequate, they should be given first consideration when a stainless steel is required.

33.   Martensitic stainless steels frequently contain significant amounts of carbon since they are used in the quenched and tempered structure. The carbon is dissolved in the austenite at elevated temperature and then trapped into the body-centered structure by quenching. Different amounts of carbon provide different levels of strength, as in the plain-carbon and alloy steels.

34.   Stainless steels are only "stainless" when there is at least 12% chromium in atomic form, free to react with oxygen at the surface. If the chromium has already reacted with another element, such as dissolved carbon (as occurs when the martensitic stainless steels are slow cooled or annealed), the material is susceptible to red rust corrosion. Martensitic stainless steels are usually not corrosion resistant when annealed, but become stainless when in the quench-and-temper condition.

35. Austenitic stainless steels are nonmagnetic and offer superior corrosion resistance to a host of media. Formability is outstanding, and they respond well to strengthening by cold work.

36. Duplex stainless steels have a chemistry and processing designed to produce a microstructure that is a combination of ferrite and austenite, and properties that are often superior to either the straight ferritic or austenitic varieties.

37. Sensitization of a stainless steel is the loss of corrosion resistance that occurs when the local concentration of chromium drops below 12%. This is usually caused by the formation of chromium carbides along grain boundaries. Methods of prevention include: keep the carbon content low, tie up the carbon with an alternative element, and rapidly cool the material through the carbide-forming temperature range.

38. Tool steels are metals designed to provide wear resistance and toughness combined with high strength. They are basically high-carbon steels where the alloy chemistry provides the desired balance of toughness and wear resistance.

39. While the AISI-SAE designation system for plain-carbon and alloy steels is based on material chemistry, the AISI-SAE system for tool steels identifies materials by a letter indicating the primary feature, such as quenching medium, primary application, special characteristic, or specific industry, followed by a number that simply designates the specific member within the family.

40. Air-hardenable tool steels can be hardened by less severe quenches, permitting tighter tolerances through heat treatment and reduced tendency to crack or warp. Applications involving large amounts of costly or precision machining are particularly attractive.

41. Hot-work tool steels generally use additions of the carbide-forming alloys, such as chromium, tungsten, and molybdenum.

42. If alloy cast irons are to be heat treated, the alloy elements are often selected to improve hardenability. If the cast iron is not to undergo heat treatment, the alloy elements are often selected to alter the properties through affecting the formation of graphite or cementite, modifying the morphology of the carbon-rich phase, or simply strengthening the matrix material. Other reasons for an alloy addition might include improving the wear resistance or providing some degree of enhanced corrosion resistance.

CASE STUDY - CHAPTER 6
Interior Tub of a Top-Loading Washing Machine

1.   The present product is currently performing in an adequate manner and has established itself as somewhat of an industry standard. The material is relatively inexpensive, and readily available, but the necessary surface treatment requires considerable energy and handling with the coating, drying and firing, often of multiple layers. The deep drawing of the material will most likely require intermediate anneals, which will further increase manufacturing cost. While both of the above areas include significant possibilities for problems and involve additional cost, it is likely that the coating process would be the most problematic and most costly.

2.   The conversion to stainless steel would enhance customer attractiveness, but also eliminate the need for a coating operation. Unfortunately, the stainless is a more costly material, would require more force to deform, and, depending on the particular type, may have poorer formability. Because of the higher forming forces, equipment and tooling would have to be stronger and would therefore be more costly.

3.   Because of the superior formability, some form of austenitic stainless steel would be preferred. This part requires conversion from flat sheet to a deep drawn shape, a procedure that will likely require multiple stages of forming. In addition, the austenitic stainlesses offer superior corrosion resistance, and the product will come into contact with a wide spectrum of water qualities, laundry products, and additives, such as chlorine bleach. There is no need for the high strength of the heat-treated martensitic grades, and the less expensive ferritic alloys lack the superior formability of the face-centered cubic structured austenitic material. Because of the spectrum of possibilities, no attempt is made to select a specific alloy.

4.   The austenitic stainless steels strengthen considerably when cold worked, and this can be a useful means of achieving the desired strength. However, it is doubtful that the additional strength of cold working is necessary, and the residual stresses imparted by the deformation may be detrimental in the form of stress-corrosion problems during service.

If the strengthening of cold work is deemed desirable, the effects of the residual stresses could be reduced by taking the material through the recovery stage of the recrystallization process. This reduces the residual stresses while retaining the mechanical properties set by the cold working.

If intermediate anneals are required during the deformation sequence, the effects of prior cold work will not be carried to the finished product. In addition, one consequence of partial cold rolling of the starting material will be to reduce the ductility of a material being used for an application that requires extensive deformation. It is unlikely, therefore, that the use of prior cold rolling would be appropriate or desirable for this product.

5.   It is possible that a surface passivation treatment would be beneficial for this product, but the inherent properties of the stainless should be adequate.

Chapter 7

NONFERROUS METALS AND ALLOYS

1.   Nonferrous metals often possess certain properties not usually associated with ferrous metals, among them being: corrosion resistance, ease of fabrication, high electrical and thermal conductivity, light weight, strength at elevated temperatures and color.

2.   The nonferrous alloys are generally inferior to steel in terms of strength and elastic modulus, and possibly weldability.

3.   Alloys with low melting points are often easy to cast, using sand molds, permanent molds, or dies.

4.   The wide use of copper and copper alloys is largely due to the high electrical and thermal conductivity, high ductility, and corrosion resistance.

5.   The relatively low strength and high ductility make copper quite attractive for forming operations. By cold-working, the tensile strength can be raised from about 30,000 psi to over 65,000 psi., with a concurrent drop in elongation from 60% to about 5%. The low recrystallization temperature is attractive when additional cold working is desired.

6.   A primary limitation of copper is its high density -- heavier than iron. Strength-to-weight comparisons place it below most engineering metals. In addition, some significant problems can occur when the metal is used at elevated temperature.

7.   In the CDA designation system, wrought alloys are assigned numbers between 100 and 799. Cast alloys have numbers between 800 and 999.

8.   The copper-zinc alpha brasses are quite ductile and formable, achieve good strength through cold working, and have good corrosion resistance and high electrical and thermal conductivities. Both strength and ductility increase with zinc content up to about 36% zinc. In addition, variations in chemistry can be used to produce changes in color and various platings are easily applied.

9.   The term "bronze" can be particularly confusing. While the term frequently refers to copper-tin alloys, it can be used to describe any copper alloy where the major alloy addition is neither zinc nor nickel.

10.   The copper-nickel alloys are particularly well known for their high thermal conductivity and high-temperature strength, coupled with good corrosion resistance.

11.   Copper-beryllium alloys can be age hardened to produce the highest strengths of the copper-based metals. In addition to having strengths similar to steel, the alloys are nonsparking, nonmagnetic, and have high electrical and thermal conductivity. Its use has been drastically limited, however, by concerns over the toxicity of the beryllium.

12.   Aluminum and its alloys have achieved popularity due to their light weight, high electrical and thermal conductivity, good corrosion resistance, and workability.

13.   The electrical conductivity of pure aluminum is approximately 62% that of copper for the same size wire and 200% that of copper on an equal weight basis.

14.   Aluminum alloys are inferior to steel in the area of elastic modulus. In addition, the wear, creep, and fatigue properties are generally rather poor.

15.   The observed corrosion resistance of aluminum alloys is again the result of a tight, adherent oxide coating, similar to that found in stainless steels.

16.   In the aluminum designation system, wrought alloys have four-digit numbers, while cast alloys have three-digit designations.

17.   While the four digit number of an aluminum alloy only designates chemistry, the temper designation or suffix denotes the condition or nature of the prior processing history of the material. This can be used to provide a good indication of the structure and properties of the alloy.

18.   The high-strength, "aircraft-quality" aluminum alloys generally receive their strengthening through an age hardening treatment.

19.   The aluminum alloys used for permanent mold casting must be designed to have lower coefficients of thermal expansion because the molds offer restraint to the dimensional changes that occur upon cooling. Die casting alloys require high degrees of fluidity and "castability" because they are often cast into thin sections. In addition, many are designed to have rather high as-cast strength under rapid cooling conditions.

20.   The aluminum-lithium alloys offer higher strength, greater stiffness and lighter weight than most of the commercial aluminum alloys, coupled with the relative ease of fabrication of aluminum alloys.

21.   Magnesium and magnesium alloys can be characterized by poor wear, creep, fatigue, and corrosion resistance properties. The modulus of elasticity is low and the alloys possess limited ductility.

22.   The use of magnesium is generally restricted to applications where light weight is very important. Magnesium alloys are
best suited for applications where lightness is the primary
consideration and strength is a secondary requirement.

23.   The forming behavior of magnesium alloys is poor at room
temperature, but most conventional processes can be performed
when the material is heated to between 450 and 700°F.

24.   Magnesium is flammable or explosive when it is in a
finely-divided form, such as powder or chips. A critical feature
here is the ratio of surface area to volume. In addition,
magnesium is flammable when heated above 800°F in the presence of
oxygen.

25.   The primary application of pure zinc is the galvanizing of
iron and steel. The principal use of the zinc-based alloys is in
die-casting operations. They are low in cost, have low melting
points, do not affect steel dies adversely, and can possess good
strength and dimensional stability.

26.   Titanium and its alloys are strong, lightweight, corrosion
resistant, and offer strengths similar to steel at temperatures
up to 900°F.

27.   The attractive mechanical properties of titanium and
titanium alloys are generally retained at temperatures up to
900°F.

28.   The nickel-based Monel alloys probably offer better
corrosion resistance to more media than any other commercial
alloy.

29.   Nickel, iron and nickel, or cobalt forms the base metal for
the superalloys.

30.   When the operating temperature exceeds the limits of the
superalloys, exotic materials must be employed, such as
TD-nickel or the refractory metals.

31.   The refractory metals consist of: niobium, molybdenum,
tantalum, rhenium and tungsten.

32.   While the eutectic lead-tin alloy offers the lowest melting
temperature of the lead-tin solders, the high cost of tin has
prompted many users to specify solders with a lower-than-optimum
tin content.

33.   Graphite possesses the unique property of actually
increasing in strength as the temperature is increased. This
makes the material attractive for elevated temperature applications, such as electrodes in furnaces.

CASE STUDY - CHAPTER 7
Nonsparking Wrench

Many safety tools have been made from the copper-2% beryllium alloy, since its age-hardened properties approach and often exceed those of many heat-treated alloy steels. This is fine for small tools where the cost of the material and the weight of the copper alloy (greater than that of steel) are not objectionable. However, with the proposed pipe wrench, both cost and weight may pose serious problems to the acceptance of the tool.

The copper-2% beryllium alloy will likely have to be used in the actual jaws of the wrench, as it is one of the few nonferrous materials that can provide the necessary strength, wear resistance, and fracture resistance for this use. However, the handle, adjuster ring, and moving L-shaped upper jaw will likely have lower mechanical property requirements that could be met by some of the other age-hardenable, higher-strength nonferrous materials. Aluminum alloys, such as 6061, 2014, 2024, 7075, 7079 and others could be forged and heat-treated to produce the handle and jaw components, and copper-beryllium inserts can be installed in the jaws. Alternately, an age-hardenable aluminum casting alloy could be selected and these components could be fabricated by sand, permanent mold, or even die casting. One problem with the use of the aluminum alloy with a copper insert would be the presence of a galvanic corrosion cell (dissimilar metals), which could be aggravated by some of the environments in which tools are typically stored. Since replaceable inserts would be desirable, and the method of assembly would likely involve a removable fastener, electrical contact between the components would be virtually assured. The possible severity of this problem would have to be monitored.

Alternative solutions would not be as attractive. Manufacture of the handle from a less expensive copper-base alloy would reduce cost and significantly reduce the galvanic corrosion problems, but the weight of such a wrench may be objectionable. Smaller wrenches in the series might be made in this manner. Magnesium alloys offer light weight, but lack the necessary strength and rigidity. Titanium alloys are difficult to fabricate (too reactive to easily cast and generally require isothermal forging). Nickel-base alloys would offer no cost advantage to the copper-beryllium.

Unless the galvanic corrosion problems become excessive in the jaws, the most attractive solution would appear to be a to use copper-beryllium inserts in cast or forged aluminum components. All parts would probably require strengthening through age hardening treatments.

Chapter 8

NONMETALLIC MATERIALS:
PLASTICS, ELASTOMERS, CERAMICS, AND COMPOSITES

1.   Some of the naturally occurring nonmetallic engineering materials are: wood, stone, clay, and leather.

2.   The term "nonmetallic engineering material" now includes plastics, elastomers, ceramics and composites.

3.   The term "plastics" refers to engineered organic materials, composed of hydrogen, oxygen, carbon and nitrogen, in the form of large molecules that are built up by joining smaller molecules. They are natural or synthetic resins, or their compounds, that can be molded, extruded, cast, or used as thin films or coatings.

4.   A saturated molecule is one to which no additional atoms can be added. If the molecule is a pure hydrocarbon, it contains the maximum number of hydrogen atoms. An unsaturated hydrocarbon does not contain the maximum number of hydrogen atoms.

5.   Polymerization can take place by either addition or condensation. In addition, a number of small molecules unite to form a large molecule with repeated units. Condensation polymerization results in the formation of a polymer and a small by-product molecule.

6.   The terms thermoplastic and thermosetting refer to a material's response to elevated temperature. Thermoplastic materials soften with increasing temperature and become harder or stronger when cooled. The cycle can be repeated as often as desired and no chemical change is involved. In the thermosetting materials, elevated temperatures tend to promote an irreversible condensation reaction. Once set, additional heatings do not produce softening. Instead, the materials maintain their mechanical properties up to the temperature at which they char or burn.

7.   When a thermoplastic material is cooled from above its melting temperature, it can assume characteristics that have been described as "rubbery", then "leathery", and finally "glasslike" as the material becomes progressively stronger and less ductile.

8.   The strength of the thermoplastic materials can be altered by mechanisms that restrict or alter the intermolecular slippage. These mechanisms include: longer chains, polymers with large side groupings, branched polymers, cross-linking, and crystallization.

9.   The deformation of thermosetting material requires the simultaneous breaking of numerous primary bonds. Therefore, these materials are strong, but brittle.

10.   Upon subsequent heating, the thermosetting polymers maintain their mechanical properties up to the temperature at which they char or burn.

11.   While thermoplastic materials are easily molded, the temperature of the mold must be cycled to permit the molded product to cool and strengthen prior to ejection. In contrast, the mold used to process thermosetting polymers can operate at a fixed temperature, but the molding time is often longer because of the need to complete the curing or "setting" of the resins.

12.   Attractive engineering properties of plastics include: light weight, corrosion resistance, electrical resistance, low thermal conductivity, the variety of optical characteristics, formability, surface finish, low cost, and low energy content.

13.   The inferior properties of plastics generally relate to mechanical strength. Yield strength, impact strength, dimensional stability, property retention at elevated temperature, sensitivity to humidity, and degradation under certain forms of radiation are all limiting or undesirable properties.

14.   Environmental conditions that may adversely affect the performance of plastics include: elevated temperature, humidity, and ultraviolet and particulate radiation.

15.   Additive agents are frequently added to plastics to improve their properties, reduce their cost, improve their moldability, and impart color.

16.   Filler materials are added to molded plastic to: improve strength, stiffness, or toughness; reduce shrinkage; reduce weight; or provide cost-saving bulk.

17.   Orienting a plastic permits superior properties to be obtained in a desired direction.

18.   The "true engineering plastics" offer improved thermal properties, first-rate impact and stress resistance, high rigidity, superior electrical characteristics, excellent processing properties, and little dimensional change with temperature or humidity. They offer a balanced set of engineering properties.

19.   Plastics have replaced glass in containers and flat glass. PVC competes with copper and brass in pipe and plumbing fittings. Plastics have replaced ceramics in sewer pipe and lavatory facilities. New automotive uses include engine components and fuel tanks and fittings.

20.   Mixed plastics contain multiple types of resins, fillers and colors, and may mix thermoplastics and thermosets. Most, however, have the same physical properties, making separation extremely difficult.

21.   Elastomers are a class of linear polymers that display an exceptionally large amount of elastic deformation when a force is applied, frequently stretching to several times their original length. In these materials, the long polymer chain is in the form of a coil, which elastically uncoils and recoils in response to loads.

22.   By cross-linking the molecules, it is possible to prevent viscous deformation, while retaining the large elastic response. The elasticity or rigidity of the product can be determined by controlling the number of cross-links. Small amounts of cross-linking produces soft, flexible material. Additional cross-linking makes the material harder, stiffer, and more brittle. Thus the properties of an elastomer can be tailored through control of the amount of cross-linking.

23.   Natural and artificial elastomers can be compounded to provide a wide range of characteristics, ranging from soft and gummy to hard, such as ebonite. These materials are outstanding for their flexibility, good electrical insulation, energy absorption and damping, durability, and resistance to many hostile environments.

24.   The outstanding physical properties of ceramics include their ability to: withstand high temperatures, provide a variety of electrical properties, and resist wear.

25.   The crystal structures of ceramic materials are frequently more complex than those for metals because atoms that differ greatly in size must be accommodated within the same structure and interstitial sites become extremely important. In addition, charge neutrality must be maintained throughout the structure of ionic materials. Covalent materials can only have a limited number of nearest neighbors - forcing inefficient packing and low density.

26.   The refractory ceramics are materials that are designed to provide acceptable mechanical and chemical properties while at high temperatures.

27.   The dominant property of the ceramic abrasives is their high hardness.

28.   Cermets are combinations of metals and ceramics that are bonded together in the same manner in which powder metallurgy parts are produced. They combine the high refractory character-istics of ceramics and the toughness and thermal shock resistance of metals.

29.   Ceramic materials generally do not exhibit their potentially high tensile strength because small pores or flaws act as stress concentrators and their effect cannot be reduced by plastic flow.

30.   The mechanical properties of ceramics generally show a wider statistical spread than the properties of metals since the size, number, shape and location of the flaws is likely to differ from part to part, inducing failure at very different applied loads.

31.   Even if all of the flaws or defects could be eliminated from the structural ceramics, the materials would still fail by brittle fracture with little, if any, prior warning. Thermal shock may be a problem, cost would be high, joining to other materials is difficult, and machining limitations favor net-shape processing.

32.   The structural ceramic materials include: silicon nitride, silicon carbide, partially stabilized zirconia, transformation-toughened zirconia, alumina, sialons, boron carbide, boron nitride, titanium diboride, and ceramic composites.

33.   Sialon is stronger than steel, extremely hard, and light as aluminum. It has good resistance to corrosion, wear and thermal shock, is an electrical insulator, and retains good tensile and compression strength up to $2550^{\circ}F$. In addition, its thermal expansion is quite low compared to steel or polymers. When overloaded, however, it will fail by brittle fracture.

34.   If perfected, a ceramic engine block would allow higher operating temperatures with a companion increase in engine efficiency, and also permit the elimination of radiators, fan belts, cooling system pumps, coolant lines, and coolant. The estimated fuel savings could amount to 30% or more.

35.   Some of the ceramic materials currently being used as cutting tools include: silicon carbide, cobalt-bonded tungsten carbide, silicon nitride, cubic boron nitride, and polycrystalline diamond. The ceramic cutting tools offer low wear rates, low friction, high rates of cutting, and long tool life.

36.   A composite material is a heterogeneous solid consisting of two or more components that are mechanically or metallurgically bonded together. Each of the components retains its identity, structure and properties, yet by combining the components, unique properties are imparted to the composite.

37.   The properties of composite materials generally depend upon: the properties of the individual materials; the relative amounts of the components; the size, shape and distribution of the discontinuous components; the degree of bonding between the components; and the orientation of the various components.

38.   The three principal geometries of composite materials are: laminar or layer-type, particulate, and fiber-reinforced.

39.   A bimetallic strip consists of two metals with different coefficients of thermal expansion bonded together as a laminate. Changes in temperature produce a change in shape.

40.   The attractive aspect of the strengthening that is induced in the dispersion-strengthened particulate composites is the stability and retention that is observed at elevated temperatures.   The particles are selected to be insoluble in the matrix material, their effect persists to temperatures much higher than for the naturally-occurring two-phase materials.

41.   Due to their unique geometry, the properties of particulate composites are usually isotropic. This is usually not true for the laminar, whose properties differ perpendicular and within the plane of the laminate. Fiber-reinforced composites, may or may not be isotropic depending on the length and randomness of the orientation of the fibers.

42.   In a fiber-reinforced composite, the matrix supports and transmits loads to the fibers, and provides the ductility and toughness. The fibers, on the other hand, provide strength by carrying most of the load.

43.   In a fiber-reinforced composite, the fibers can be in a variety of orientations: short, random fibers; unidirectional fibers; woven fabric layers; and complex 3-dimensional weaves.

44.   The properties of fiber-reinforced composites depend strongly upon: the properties of the fiber material, the volume fraction of fibers, the aspect ratio of the fibers, the orientation of the fibers, the degree of bonding between the fiber and the matrix, and the properties of the matrix.

45.   Compared to metals, the metal-matrix composites offer higher stiffness and strength and a lower coefficient of thermal expansion. Compared to the organic matrix materials, they offer higher heat resistance as well as improved electrical and thermal conductivity.

46.   In a ceramic matrix composite, the fibers add directional strength, increase fracture toughness, and improve thermal shock resistance.

47.   Current limitations to the extensive use of composite materials in engineering applications include: the high cost of the material, the intensity of labor required for fabrication, and the lack of trained designers, established design guidelines, information about fabrication costs, and methods of quality control and inspection. In addition, it is often difficult to predict interfacial bond strength, strength of the composite, response to impacts and probable modes of failure. There is concern about heat resistance, sensitivity to various environments, and instability of properties. Repair, maintenance, and assembly are difficult or require special procedures.

48.   Composites are quite attractive for aerospace applications because they offer high strength, light weight, high stiffness, and good fatigue resistance.

PROBLEMS FOR CHAPTER 8

1. a). Some of the desirable features for a submarine material are high strength (to withstand water pressure), fracture resistance (to withstand possible impacts), corrosion resistance (to both fresh and salt water), and the ability to be fabricated into a leak-tight assembly (possibly using techniques like welding). Possible materials would include high-strength steels, titanium alloys, and possibly nickel-based alloys.

   b). For aerospace applications, concerns focus on areas such as: light weight, strength-to-weight ratio, fatigue resistance, and ease of fabrication and ability to fabricate in small production quantities.

   c). Engineers are constantly pushing the limits of engineering materials. Some current targets that are presently unattainable include: (1) reuseable rocket engines that can withstand temperatures in excess of 4000^OF, stresses and severe vibrations, and (2) light weight wing skin materials for hypersonic aircraft that will withstand temperatures in excess of 1800^OF, and be resistant to fracture, fatigue and corrosion.

   Since these applications both require elevated-temperature properties, they will likely be addressed through ceramic materials, the family of intermetallic compounds, or even the high-temperature metals (although these are sufficiently heavy as to be inappropriate for the airplane use). Any use of polymers would be highly unlikely at the specified temperatures.

2.   This is an open-ended problem, but numerous examples can be considered, such as: (1) window frames (wood, metal, vinyl); (2) lavatory basins (metal, cast polymer, ceramic whiteware); and (3) window cranks for autos or mobile homes (die cast plastic or die-cast zinc).

CASE STUDY - CHAPTER 8
Two-wheel Dolly Handles

   There is considerable variability to this problem, but concerns should address the durability, impact resistance when dropped on the handles, and the ability to withstand high localized stresses (such as at the bolt holes) without cracking or fracture. Since the product may be used on outdoor loading docks in mid-winter, critical properties must be present at low temperatures -- a possible problem for polymeric materials.

A number of alternatives are possible, including such techniques as the injection molding of polymeric material containing chopped fibers, and others. Since the design was made for casting, one might expect incorporation of pattern-removal draft, and a preference for uniform thickness or section size.

Chapter 9

MATERIAL SELECTION

1.   In a manufacturing environment, the selection and use of engineering materials should be a matter of constant reevaluation. New materials are continually being developed. Others may no longer be available. Prices are subject to change and fluctuation. Concerns regarding environmental pollution, recycling, and worker health and safety impose new constraints. The desire for weight reduction, energy savings, or improved corrosion resistance may require a material change. Increased competition, the demand for improved quality and serviceability, and negative customer feedback may all prompt review and evaluation. Finally, the climate of product liability demands constant concern for engineering materials.

2.   Recent shifts in the materials used in automobiles show increased use of lighter weight materials and high-strength steels, as well as plastics and composites. Early automobiles made extensive use of wood, and some early fenders were made of leather.

3.   Aerospace demands for lighter weight, higher strength and higher stiffness have prompted considerable development and use of advanced composites. The desire for higher speeds will require light weight materials capable of withstanding high elevated temperatures.

4.   There is a distinct interdependence between engineering materials and the processes used to produce the desired shape and properties. A change in materials will often require a change in manufacturing processes; and improvements in processes may lead to a reevaluation of materials.

5.   The three usual phases of product design are: conceptual design, functional design and production design. Consideration of material is of almost no concern in the first phase; is of importance in the second phase in that suitable materials must be available and selected; and in the third phase, the exact materials to be used must be related to the production processes and to the tolerances required and the cost.

6.   If one does not require that prototype products be manufactured from the same materials that will be used in production and by the same manufacturing techniques, it is possible to produce a perfectly functioning prototype that cannot be manufactured economically in the desired volume or one that is substantially different from what the production units will be like. By using the same material and process, the prototype will provide a true assessment of the performance and manufacturability of the product.

7. New materials should be evaluated very carefully to assure
that all of their characteristics are well established. Numerous
product failures have resulted from new materials being
substituted before their long-term properties were fully known.
When changing a process, it is important that the effect of the
process on the properties of the material be known and
acceptable.

8. The "case-history" approach has several pitfalls. First,
minor variations in service requirements may well require
different materials or different manufacturing operations. In
addition, this approach precludes the use of new technology, new
materials, and other manufacturing advances that may have
occurred since the formulation of the previous solution.

9. The most frequent problem that arises when seeking to
improve an existing product is to lose sight of one of the
original design requirements and recommend a change that in some
way compromises the total performance of the product.

10. A thorough job of defining needs is the first step in any
materials selection, and all factors and possible service
conditions should be considered. Many failures and product
liability claims have resulted from simple engineering oversights
or failure to consider all types of reasonable product use.

11. The compatibility of a product to its service environment
is absolutely necessary for its success. Some considerations
should include: highest, lowest, and normal operating temperatures, and the nature of any temperature changes; possible
corrosive environments; desired lifetime; and the anticipated
level of maintenance or service.

12. Although there is a tendency to want to jump to "the
answer", it is important that all factors be listed and all
service conditions and uses be considered. Many failures and
product liability claims have resulted from simple engineering
oversights or the designer not anticipating reasonable use for a
product or conditions outside of the specific function for which
he designed it.

13. "Absolute" requirements are those for which no compromise
or substitution can be permitted. "Relative" requirements are
those which can be compromised to some extent.

14. Handbook-type data is obtained through the use of
standardized materials characterization tests. The conditions of
these tests may not match with those of the proposed application.
Significant variation in factors such as temperature, rates of
loading, and surface finish can lead to major changes in material
performance. In addition, the handbook values often represent an
average, and the actual material properties will vary on either
side of that value.

15.   While cost is indeed an important consideration, it may be desirable to first demonstrate that the material or materials meet all of the necessary requirements. If more than one candidate emerges, then cost becomes a factor. If only one is satisfactory, then one must determine if its cost is acceptable.

16.   Barbell weights should be evaluated on a cost per pound basis. Parts with fixed size, like door knobs should be evaluated on a cost per cubic inch basis. There are numerous other examples.

17.   Product failures can provide valuable information. By identifying the cause of the failure, the engineer can determine the necessary changes that would be required to prevent future occurrences. Failures of similar parts in similar applications can provide additional information.

18.   In selecting a material, one should consider the possible fabrication processes and the suitability of the various candidate materials to each process. All processes are not compatible with all materials. The goal is to arrive at the best combination of material and manufacturing process for the particular product.

19.   Caution should be exercised when substituting a new material in an existing design because it is possible to cause more harm than good if certain features are overlooked. The new material generally will not possess all of the characteristics of the original one - if some of the absent properties are necessary, the component will fail. A materials substitution should be approached as thoroughly as a new product.

20.   Product liability cases have resulted from a number of reasons, most commonly: failure to know and use the latest information about the material used, failure to foresee and take into account reasonable uses for the product, use of materials about which insufficient data is known, inadequate quality control, and material selection made by unqualified people.

21.   A single-page comparative rating chart is a useful tool in material selection. Various properties can be weighted as to significance, and candidate materials can be evaluated in an equal and identical fashion. In addition, it is less likely that an important property will be overlooked.

1. Based on the chart below, material Y has the highest rating number. However, because it does not have satisfactory weldability and this is an "absolute" requirement, it should not be selected. Material Z should be used.

Material	Go-No-go Screening			Relative Rating Number								Material Rating Number
						(1)	(4)	(5)		(4)		
	Corrosion	Weldability	Brazability	Strength	Toughness	Stiffness	Stability	Fatigue	As-welded Strength	Tensile Strength	Cost	
X		S				3x1	3x4	2x5		3x4		37
Y		U				3x1	5x4	3x5		5x4		58
Z		S				3x1	3x4	5x5		2x4		48

CASE STUDY - CHAPTER 9
Material Selection

This is an open-ended and extremely variable problem that is designed to get the student to question why parts are made from a particular material and how they could be fabricated to their final shape. In addition, they are asked to consider the need for property modification via heat treatment and/or surface treatment, and should begin to recognize the need to properly integrate these operations in the manufacturing sequence. The specific answers received will depend not only upon the specific product or products chosen but also upon the background and perception of the student.

Chapter 10

MEASUREMENT AND INSPECTION

1.   In order for parts to be interchangeable, they must be manufactured to the same standards of measurement. Simply put, everybody's definition of an inch or a centimeter must be the same identical measurement. In addition, certain sizes and shapes (like threads on a shaft or teeth on a thread) are standardized. Thus, all spark plugs for automobile engines have a standard diameter size and thread shape to fit into everyone's sockets. Standardization is fundamental to interchangeability and interchangeability is fundamental to repetitive part manufacture and mass production.

2.   The least expensive time to make a change in the design is before the part is being made. Putting the manufacturing engineering requirements into the design phase helps insure that the part can be economically fabricated.

3.   Attributes inspection trys to determine if the part is good or bad. Variables inspection requires a measurement be made to determine how good or how bad and thus, more information about part quality is obtained. If your car has an oil pressure gage, you always know what the oil pressure is (variables), but if it only has a warning light, you only know whether the pressure is good (no light) or bad (light comes on).

4.   Warning lights (usually red) readily alert the driver to a bad situation, whereas the driver may completely ignore a low gage reading. The driver may not even know what a bad reading is or that a dangerous condition exists, or worse, what the gage is actually informing him or her about. Most cars today have both kinds of inspection devices to keep the driver informed. Sometimes the decision to change is based on economics as attributes gages are usually less expensive than variables types.

5.   The four basic measures are: length, time, mass, and temperature.

6.   Referring to Figure 10-1, the Pascal is a measure of pressure in SI units. Pressure is the force per unit area and its dimensions are newtons per square meter (N/m^2) in SI units or psi in English units. This unit is named after Blaise Pascal (1623-1662), a French mathematician and scientist who developed the following principal - a pressure applied in any portion of the surface of a confined fluid is transmitted undiminished to all points within the fluid - Pascal's principal.

7.   The grades of gage blocks are laboratory, precision, and working - in decreasing level of accuracy. The blocks come in sets so that they can be "wrung" together into any length needed from 0.1001 to over 25 inches in increments of 0.0001 inch.

8.   The surface tension of an ultrathin film of oil between the
very smooth, flat, block faces keeps the blocks locked together.
Because they are so smooth and in such intimate contact, they can
actually weld together via diffusion if left in contact for
prolonged periods of time.

9.   The allowance determines the desired basic fit between
mating parts. Tolerance takes into account deviations from a
desired dimension and fit, and are necessary in order to make
manufacturing practicable and economical.

10.   (a) Sliding fit would be too loose and wring fit, too
tight - therefore, snug fit, hand assembled. (b) Obviously, a
sliding fit as the speed is very low. (c) Free fit with
liberal allowance as speeds are high and so are pressures.

11.   Most of the examples given here by the students will come
from the suggestions in the book -- attaching wheels to shafts,
like a metal railroad wheel to the axle. The parts suggested
here should be metal, in the main, as most medium force fits are
between metal parts.

12.   A shrink fit is permanent, but can be disassembled by
proper heating and/or cooling of the members. The word shrink
implies that one element is heated (to expand it) and the other
is cooled (to shrink it). Then the elements are joined to form a
shrink fit. A weld is absolutely permanent -- cannot be
disassembled without ruining the parts.

13.   To determine the aim of a process, one needs measures of
accuracy. To determine the variability in a process, one needs
measures of precision. Accuracy is measured by distribution
means and precision is measured by variances or standard
deviations (square roots of variances). A process capability
study is usually performed by taking samples of the output from
the process and measuring them for the desired characteristic.

14.   Interferometry is an example of an optical inspection
method.

15.   The factors should include the rule-of-ten, linearity,
repeat accuracy, stability, resolution and magnification, the
type of device, the kind of information desired (attributes or
variables), the size of the items to be measured, the rate at
which they must be measured, and the economics of buying,
installing, and using the device.

16.   Determining repeat accuracy is easy. Just step on the
scale and step off numerous times and take readings. Determining
linearity requires that you have a set of standard weights which
you can load on and off -- say 10, 50, 90, 130 etc. pounds -- and
plot linear loads versus readings. Generally, scales and other
measuring devices are nonlinear at the ends of the scale and
linear in the central area.

17.   Variable with student. The experiment should show that magnification amplifies the measurement while resolution refers to the limit of detection.

18.   Magnification of the output of a measuring device beyond the limits of its resolving capability is of no value. Magnification of a photographic negative beyond the size of the silver halide grains results in grainy photographs. Every measuring device has a limit to its resolving capability. All the magnification in the world will not change that limit.

19.   Parallax is the apparent change in the position of an object resulting from the change in the direction or the position from which the object is viewed. This is why tennis linesmen sit looking directly down a line during a match and try to keep their heads still. A spectator sitting perpendicular to the line will see the ball differently than the linesman.

20.   The measuring instrument should be an order of magnitude (10 times) more precise than the object being measured. This rule actually refers to the gage capability. Gage capability is determined by gage R&R studies. See Statistical Quality Design and Control by DeVor, et al.

21.   The 25 divisions of the moveable vernier plate are equal in length to the 24 divisions on the main scale. Thus each division on the vernier equals 1/25 of .6 or .024 inches. Each division on the main scale is equal to 1/24 of an inch or .6 or 0.025 inches. Thus each division on the vernier is 0.025 - 0.024 = 0.001 inches less than each division on the main scale.

22.   The micrometer is sensitive to the closing pressure and the lack of pressure control. Errors in analogue devices are also made by misreading the barrel by a factor of 0.025.

23.   They are both about the same order of magnitude in terms of their precision and repeatability, but the micrometer has a limited size range and, thus, must be purchased in sets (quite expensive), whereas a vernier can measure a wide range of sizes with one device. The micrometer is more rugged and better suited for the industrial setting (shop floor). It is also less sensitive to dirt and it is easier to teach someone how to read it.

24.   The device tends to lift itself off the surface if too much torque is applied.

25. The equation for thermal expansion is $\Delta l = \alpha\, l\, \Delta T$ where Δl is the change in length for a given change in temperature (ΔT). α is the coefficient of thermal expansion ($11 \times 10^{-6}/\,^{\circ}C$) and l is the length of the bar (2 feet). $20\,^{\circ}F = 6.67\,^{\circ}C$ and 2 ft = 24 inches. Therefore:

$$\Delta l = 11 \times 10^{-6} \times 24 \times 6.67 = 0.017768 \text{ in.}$$

which is well within the measuring capability of a supermicrometer. However, don't forget that the supermicrometer will also expand (or contract) with this temperature change, so if you tried this experiment, you would not get this reading unless only the steel bar expanded, not the supermicrometer itself. You can detect a change in length of a bar with a supermicrometer simply due to heating with your hands.

26. Optical means are used so that nothing touches and thus distorts a delicate part.

27. Parts can be measured directly using the micrometer dials or compared with a profile or template drawn directly on the screen. The images on the screen can also be directly measured by a ruler and these dimensions then divided by the magnification being used -- usually 10 to 20X. The projector magnification should be checked, however, when this technique is used by projecting a known standard onto the screen.

28. Because of the large distance and the accuracy and precision needed, a laser interferometer would probably be most suitable.

29. The laser scanner is more precise and likely to be faster with less image processing.

30. The CMM is a mechanical device with precise X - Y - Z movements for precision 3D measurements. Usually a probe is used to touch the surfaces of parts being measured and the dimensions are read on digital displays and computer terminals.

31. A sine bar uses the principle that the sine of a right triangle is the ratio of the side opposite the angle to the hypotenuse.

32. The not-go member is usually made shorter than the go member because it undergoes less wear.

33. In using a dial gage, one must be sure that the axis of the spindle is parallel with the dimension being measured. Dial indicators also suffer from friction in the gears, so multiple readings are highly recommended.

34. The gage is designed so that if it errors, it will reject a good part rather than accept a bad part. The gage has a tolerance added for manufacturing and a tolerance added for wear.

35. The go ring should slip over the shaft. If it does not, the shaft is too large. The not-go ring should not slip over the shaft. If it does, the shaft is too small.

36. Air gages will detect both linear size deviation and out-of-round conditions of holes. They are fast and there is virtually no wear on the gage or part.

37. Monochromatic light waves will interfere with each other (producing light and dark bands) if they get out of phase. Thus, a dark band indicates that the two beams have cancelled each other out. Light from a single source can be shifted out of phase by having it travel different distances.

38. An optical flat is made from glass or quartz, is transparent, and the two faces are flat and parallel to a high degree of accuracy. A toolmaker's flat is made of steel, with the two faces very flat, but they do not have to be exactly parallel.

39. The various machining processes will create widely different patterns in the surfaces (lay patterns). Surface roughness, as measured by stylus tracer devices (in AA or rms), reflect only an average value and very different lay patterns can have the same value of roughness.

40. Because identical roughness values can be very different in appearance, surface-finish blocks enable a designer to better relate a desired surface, obtained by a specific process, to the measured value that must be specified.

41. The spherical radius of the tip of a diamond stylus limits the resolution. Suppose you have a smooth plate with small holes on it. In your hand you have a needle and are trying to locate the holes. Let's assume the holes are square, round, and triangular in shape. The needle will allow you to detect the location of the holes, but not identify their shape when the holes are about the same size as the tip of the needle or smaller. Thus, there is a big difference between being able to detect the presence of a flaw and being able to resolve its geometry.

42. As the surface finish improves (surface gets smoother and AA or rms values get smaller), the tolerance generally improves --gets smaller. Improving the surface finish and tolerance usually means identifying better, more precise processes, so the cost goes up accordingly. The exception is finishing processes which are used to improve the surface finish without strong regard to the tolerance.

43. Devices that use light scattering correlated to surface roughness lose their validity when surface roughness gets much above 40 - 50 min. AA.

1.   Reading 1.436 in.

Inches are numbered in sequence over the full range of the bar. Every fourth graduation between the inch lines is numbered and equals one-tenth of an inch or 0.100". Each bar graduation is one twenty-fifth of an inch or 0.025".

The vernier plate is graduated in 25 parts, each representing 0.001". Every fifth line is numbered - 5, 10, 15, 20, 25 - for easy counting.

To read the gage, first count how many inches, tenths (0.100") and twenty-fifths (0.025") lie between the zero line on the bar and the zero line on the vernier plate and add them.

Then count the number of graduations on the vernier plate from its zero line to the line that coincides with a line on the bar.

Multiply the number of vernier plate graduations you counted times 0.001" and add this figure to the number of inches, tenths and twenty-fifths you counted on the bar. This is your total reading. The vernier plate zero line is the one inch (1.000") plus four tenths (0.400") plus one twenty-fifth (0.025") beyond the zero line on the bar, or 1.425". The 11th graduation on the vernier plate coincides with a line on the bar (as indicated by stars). 11 x .001" (.011") is therefore added to the 1.425 bar reading, and the total reading is 1.436".

2.   Reading 41.68 mm

Each bar graduation is 0.5 mm. Every twentieth graduation is numbered in sequence - 10 mm, 20 mm, 30 mm, 40 mm, etc. - over the full range of the bar. This provides for direct reading in millimeters.

The vernier plate is graduated in 25 parts, each representing 0.02 mm. Every fifth line is numbered in sequence - 0.10 mm, 0.20 mm, 0.30 mm, 0.40 mm, 0.50 mm - providing for direct reading in hundredths of a millimeter.

To read the gage, first count how many mm lie between the zero line on the bar and the zero line on the vernier plate.

Then find the graduation on the vernier plate that coincides with a line on the bar and note its value in hundredths of a mm.

Add the vernier plate reading in hundredths of a mm to the number of mm you counted on the bar. This is your total reading. The vernier plate zero line is 41.5 mm beyond the zero line on the bar, and the 0.18 mm graduation on the vernier plate coincides with a line on the bar (as indicated by stars.) 0.18 is therefore added to the 41.5 mm bar reading, and the total reading is 41.68 mm.

3.

      41.68 mm = 1.6409 inches
      1.6409 - 1.436 = 0.2049 inches

4.

      Sin θ = 3.250 / 5.000 = 0.65
      θ = 40.54 degrees

5.   The error due to the gage blocks will be covered up by the dial indicator error

      +0.000,008 or -0.000,004 for gage blocks
                versus
      +0.001 or -0.001 for dial indicator

The error will be 3.249 to 3.251 due to leveling of the part with the dial gage.
      θ = 40.53 to 40.55

Error   ≈.02 degrees, due to dial indicator not the gage blocks.

6.   A 0.359   B 0.242   C   0.376

7.   A   0.2991   B   0.3001

8.   Metric vernier micrometers are used like those graduated in hundredths of a millimeter (0.01 mm), except that an additional reading in two-hundredths of a millimeter (0.002 mm) is obtained from a vernier scale on the sleeve.

The vernier consists of five divisions each of which equals one-fifth of a thimble division - 1/5 of 0.01 mm or 0.002 mm.

To read the micrometer, obtain a reading to 0.01 mm. Then see which line on the vernier coincides with a line on the thimble. If it is the line marked 2, add 0.002 mm; if it is the line marked 4, add 0.004 mm, etc.

The left side micrometer reads 5.500 mm

The 5 mm sleeve graduation is visible 5.000 mm

The 0.5mm line on the sleeve is
   visible.......................... 0.500 mm

Line 0 on the thimble coincides with
   the reading line on the sleeve ... 0.000 mm

The 0 line on the vernier coincide
   with lines on the thimble........ <u>0.000 mm</u>

The micrometer reading is .... 5.500 mm

The right side micrometer reads 5.508 mm

The 5 mm sleeve graduation is visible 5.000 mm

The 0.5mm lines on the sleeve is
  visible........................... 0.500 mm

Line 0 on the thimble lies below the reading
line on the sleeve, indicating that a vernier
reading must be added.

Line 8 on the vernier coincides with
  a line on the thimble............ 0.008 mm.

The micrometer reading is......... 5.508 mm

9.
Height difference = 5 x 0.000,001,6 x 2 = 0.000,116 inches.

10.   The probe is used to find the stack of gage blocks that
exactly matches the height of the ball sitting on the part. The
angle theta is 17.354 degrees, from

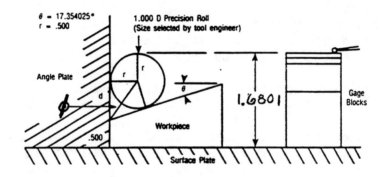

Tan θ = $\dfrac{1.125 - 0.800}{2.000}$ = 0.3125

X = height of gage blocks = .500 + d + r

  = .500 + .5 cotϕ + 0.500

Cot ϕ = $\dfrac{X - 0.500 - 0.500}{0.500}$ from the Figure

The relationship between θ and ϕ is

    90° = 2ϕ + θ , so ϕ = (90 - θ)/2 = 36.322

So Cot ϕ = 1.3602 = X - 1.0 / 0.5

X = 1.68

Chapter 11

NONDESTRUCTIVE INSPECTION AND TESTING

1.   When destructive testing is employed, statistical methods must be used to determine the probability that if certain test specimens are good, then the entire production run will be good also. Each of the products tested is destroyed during the evaluation, so the cost of such methods must be borne by the remaining products. There always remains some degree of uncertainty about the quality of the remaining products because they have never been individually evaluated.

2.   In a proof test, a product is subjected to loads of a determined magnitude, generally equal to or greater than the designed capacity. If the part remains intact, then there is reason to believe that it will perform adequately in the absence of abuse or loads in excess of its rated level.

3.   Hardness tests can be used to provide reasonable assurance that the proper material and heat treatment were employed in a given part. The tests can be performed quickly, possibly on every product, and the associated mark can easily be concealed or removed.

4.   Nondestructive testing is the examination of a product in a manner that will not render it useless for future service. The testing can be performed directly on production items or even parts in service. The entire production lot can be inspected, different tests can be applied to the same item, and the same test can be repeated on the same specimen if desired. Little or no specimen preparation is required and the equipment is often portable.

5.   Some possible objectives of nondestructive testing include: the detection of internal or surface flaws, the measurement of dimensions, the determination of a material's structure or chemistry, or the evaluation of a material's mechanical or physical properties.

6.   When selecting a nondestructive testing method, one should consider the advantages and limitations of the various techniques. Some can be performed on only certain types of materials. Each is limited in the type, size, and orientation of the flaws that it can detect. Various degrees of accessibility may be required and there may be geometric restrictions as to part size or complexity. Availability of equipment, the cost of operation, the need for a skilled operator, and the availability of a permanent record are other considerations.

7.   By ensuring product reliability and customer satisfaction, nondestructive testing can actually be an asset, expanding sales and profitability. In addition, it can be used to assist product development and process control, further reducing costs.

8.   Visual inspections are limited to the accessible surfaces of a product, so no information is provided relating to the interior structure or soundness.

9.   Liquid penetrant testing can be used to detect any type of open surface defect in metals and other nonporous materials. Cracks, laps, seams, lack of bonding, pinholes, gouges, and tool marks can all be detected.

10.   Materials must be ferromagnetic in order to be examined by the magnetic particle technique. Nonferrous metals, ceramics and polymers cannot be inspected.

11.   The relative orientation of a flaw and magnetic field is quite important in determining whether the flaw will be detected, since the flaw must produce a significant disturbance to the magnetic field. If a steel bar is placed inside an energized coil, a magnetic field is produced that aligns with the axis of the bar. Defects perpendicular to this axis can be easily revealed, but a flaw parallel to the axis could go relatively unnoticed. By passing a current through the bar, a circumferential magnetic field is produced that will detect axial flaws, but not those in a radial orientation.

12.   "Sonic testing", where one listens for the characteristic "ring" to a product, is limited to the detection of large defects because the wavelength of audible sound is rather large compared to the size of most defects.

13.   Ultrasonic inspection can reveal most internal defects, and indicate both flaw size and location.

14.   The three types of ultrasonic inspections are: (1) pulse-echo, where inspection is made from one side or surface; (2) through-transmission, where the sending and receiving transducers are on opposite sides of the piece, and (3) resonance testing, where thickness can be determined from a single side.

15.   X-rays, gamma rays and neutron beams can all be used to provide radiographic inspection of manufactured products.

16.   A penetrameter is a standard test piece that provided a reference for the image densities on a radiograph. Penetrameters are made of the same or similar material as the specimen and contain structural features of known dimensions. The image of the penetrameter then permits direct comparison with the features in the image of the product.

17.   Since the materials examined by eddy-current inspection must be good electrical conductors, it is unlikely that the technique would be useful to examine ceramic or polymeric materials.

18.   Eddy current testing can be used to detect surface and near-surface flaws, such as cracks, voids, inclusions and seams. Differences in metal chemistry or heat treatment will affect the magnetic permeability and conductivity of a metal, and hence, the eddy current characteristics. Material mix-ups can be detected. Specimens can be sorted by hardness, case depth, residual stresses, or any other structure-related property. Thicknesses or variation in thickness of platings, coatings, or even corrosion can be measured.

19.   Acoustic emission is not a means of detecting an existing, but static defect in a product, but a means of detecting a dynamic change, such as the formation or growth of a crack or defect or the onset of plastic deformation. The sound waves emitted during this dynamic event are detected and interpreted.

20.   By using multiple sensors and timing techniques similar to those used to locate earthquakes, acoustic emission can be used to physically locate the flaw or defect emitting the sound.

21.   Various thermal methods can be used to reveal the presence of defects. Parts can be heated and means used to detect abnormal temperature distributions, indicative of faults or flaws. The presence of "hot spots" on an operating component can be an indication of defects. Thermal anomalies can also provide an indication of poor bonding in composite materials.

22.   Evaluations of resistivity from one sample to another can be used for alloy identification, flaw detection, or the assurance of proper processing - such as heat treatment, the amount of cold work, weld integrity, or the depth of case hardening.

23.   Computed tomography provided a cross-sectional view of the object along the axis of inspection. By multiple scanning, full 3-dimensional representations of the interior of a product can be generated.

24.   The basis of inspection was once the rejection of any product shown to contain a flaw or defect. With the rapid advances in inspection capability, it is now possible to detect "defects" in almost every product, including those that perform adequately. The basis of discrimination should be the separation of products with critical flaws that could lead to failure, from products where the flaws will remain dormant throughout the lifetime of the product, i.e. allowable flaws.

PROBLEMS FROM CHAPTER 11

1.   X-ray radiography is a poor means of detecting cracks in a product, for the crack or void size must be sufficiently large as to produce a difference in transmitted intensity. Only if the orientation of a crack were parallel to the X-ray beam would there likely be sufficient differences to detect the flaw. Otherwise, the X-ray would indicate a given and constant thickness of material and reveal none of the existing cracks.

       Crack detection would be better performed by ultrasonic inspection, penetrant testing, and magnetic particle inspection. These have limitations as to the location, depth and orientation, but are generally superior for detecting cracks than X-ray radiography.

       For a permanent record, the electronic signals received at the transducers in ultrasonic inspection, and displayed on some form of screen, can also be recorded on magnetic tape or some other form of storage medium. The surfaces of parts examined by penetrant inspection or magnetic particle techniques can be photographed or recorded on some form of video recorder.

2.   A major limitation to each of the following is:

     - Visual inspection: Depends upon the skill of an inspector and is limited to surface flaws.
     - Liquid penetrant inspection: Can only detect flaws that are open to the surface.
     - Magnetic particle inspection: Orientation of the flaw and field affects sensitivity, limited to ferromagnetic materials, detects only surface and near-surface flaws.
     - Ultrasonic inspection: Difficult to use with complex shape parts, trained technicians are required, and the area of inspection is small.
     - Radiography: Costly, must observe radiation precautions, defects must be larger than a minimum size, must generally process film to get results.
     - Eddy current testing: Reference standards are needed for comparison and trained operators are required, materials must be conductive, depth is limited.
     - Acoustic emission monitoring: Only growing flaws can be detected, experience is required, and there is no indication of the size or shape of the defect.

3. a). Ceramic materials can be inspected by several of the traditional methods, but are limited by their poor electrical conductivity and lack of ferromagnetism. Acceptable techniques include: visual inspection, liquid penetrant, ultrasonics, and radiography.

b). Polymeric materials have similar limitations, coupled with a relatively low density that might make radiography more difficult. Acceptable techniques include: visual inspection, liquid penetrant, and ultrasonics.

c). With the composite materials, how you look often depends upon what you are trying to see. For example, is it the quality of the matrix, distribution of reinforcement, or integrity of the bond between the matrix and the reinforcement? For the most part, the technique must be compatible with the matrix material, but the effect of the reinforcement should be considered in the selection and specification.

4. High density powder metallurgy parts can be treated as conventional parts from the viewpoint of both inspection and secondary processing. As the density decreases (i.e. the volume fraction of voids increases), the material is less capable of transmitting sound, current, and magnetic field -- the essence of the various probing techniques. Moreover, the voids are actually "defects" and are often detected. When the numbers become great, the signals become quite garbled, and the presence of additional, more-significant, defects may be difficult, or impossible.

CASE STUDY - CHAPTER 11
Portable Failure Analysis Kit

1). The primary purpose of the requested kit is to collect information and specimens for examination back at the laboratory. It would be inappropriate for the investigator to try to perform laboratory-type examination in the field, especially if these could be better or more accurately performed in the lab. NOTE: On rare occasions, the mere size of the piece in question (such as a bridge) or the desire for immediate results renders such testing desirable, but such is not usually the case. Therefore, a proposed kit would contain:

a). Equipment for gathering information, such as the observations and opinions of operating personnel, supervisors, observers, etc. This would include: a notebook, pens and pencils, and a small portable tape recorder (with tapes).

b). Photographic equipment to record the failure site, the locational relationship of failed components, the stages and sequence of disassembly, and/or the location and orientation of samples removed from the failure or the failure site. A Polaroid camera with color film would assure the access of acceptable photos. A close-up lens and flash attachment would be desirable, as would an ample supply of film (a type that provides a negative would be preferred) and spare batteries for the camera. A ruler or other well known object can be included in photos to reveal the relative size of components.

c). A variety of hand-operated tools to assist in the removal of components and the collection of laboratory specimens. This might include: A hacksaw and blades, hammer, pliers, screwdrivers, knife, wrenches (socket and straight), clamps, chisels, scissors, tweezers.

d). Examination aids, such as: a flashlight with spare batteries, magnifying glass or jeweler's eyepiece (10X), low-power binocular microscope, small handheld and dental mirror.

e). Measuring devices, such as a ruler, measuring tape, micrometer, and vernier.

f). Marking devices to label specimens and denote in photos the locations of cuts or the orientation of the pieces: magic markers, chalk, grease pencils, and pens.

g). Equipment to perform several basic tests:
   - A portable dye penetrant testing kit to reveal the presence of cracks.
   - A set of triangular metal files can be tempered in the laboratory to various hardnesses to provide an inexpensive means of getting "ballpark" hardness in the field.

h). A portable drill with attachments can be used to obtain borings, wire brush surfaces, grind surfaces, produce a crude spark test, etc.

i). Equipment for identifying, preserving, and transporting specimens back to the laboratory: envelopes, labels, plastic bottles, zip-lock bags, cellophane tape and masking tape (can be used to remove and preserve corrosion products for X-ray analysis, as well as more normal uses).

j). Cleaning agents, chemical reagents (solvents, etchants, macroetches, etc.), abrasives (sand paper, toothpaste, etc.), cloths and rags, toothbrush, other small brushes.

k). Wax or clear nail-polish to coat and preserve critical fracture surfaces.

l). A small magnet - to check materials and discern various types of stainless steels.

m). Gloves and safety glasses.

n). Environmental evaluation devices: thermometer, hygrometer (humidity), litmus paper (possibly graded by pH).

o). A small vise.

p). A small propane torch - to heat or loosen components. (NOTE: be sure exposure to heat will not alter or obliterate evidence.)

q). A cold-mount kit to permanently mount small or fragile specimens.

r). Selected reference manuals on engineering materials and their properties.

2). With additional funds, one might want to consider:

a). An additional camera, such as a high-quality 35-mm with a variety of special lenses (telephoto, close-up, wide angle), and upgrade facilities for the previous Polaroid camera, which would be retained to assure the acquisition of acceptable photos.

b). A calibrated portable hardness tester, such as a portable Brinell tester.

c). An improved microscope with special eyepieces to measure case depth, thickness of plating layers, etc.

d). Portable metal identification kit (to identify specific metals and alloys).

Chapter 12

PROCESS CAPABILITY AND QUALITY CONTROL

1.   A PC study examines the output from a process in order to
determine the capability of the process in terms of its accuracy
(its aiming ability or its ability to hit the desired nominal
value) and its precision (the ability of the process to repeat
the variability in the process). Accuracy refers to the
centering of the process and variability refers to the scattering
of the values about the center value. Accuracy is measured by
the mean of the values and variability is measured by the
standard deviation of the values about the mean -- also called
the spread of the distribution.

2.   Every process has some inherent variability. The causes of
this variability may be known (assignable) but not removable (you
know what is causing the variation but it is not feasible or too
costly to remove it) or unassignable (i.e., inherent in the
process and thus not removable). The latter is its nature.

3.   It would look like Figure 12-1a with occasional holes
scattered to the right of center at various distances from the
center -- a random pattern because the wind is gusting, not
steady. In real processes, intermittent changes of this sort are
extremely difficult to isolate and identify and therefore remove
from the process as an assignable cause. This is an example of
an assignable cause that would be difficult to remove -- how do
you make the wind stop gusting without great cost (i.e. enclose
the shooting range).

4.   A good way to get students to review the steps in a P.C.
study is to have them try to do one themselves. The example of
shooting a gun given in the text can be "simulated" by having the
students work with file cards for targets and darts for the
bullets. Give each student in your class a different distance to
stand from the cards (mounted on a dart board) when the class
data is examined as a whole, you will observe the increase in
process variability with distance from the target. Depending on
the dart throwing ability of the students, there will also be a
loss of accuracy.

5.   Two "identical" machine tools doing exactly the same
process will have different amounts of process variability. The
individual machines will have different variability when the work
material is changed, the operator is changed, the specific
process on the machine is changed, etc. Thus, it is necessary to
gather data on the specific machine tool during the process
itself.

6.   Taguchi experiments can be used to determine the process capability of a process. Taguchi methods used truncated (simplified) experimental designs in which all the causes of variability are explored. They permit the variability to be reduced by selecting the proper combination of input variables to reduce the noise (i.e., the variability) in the output.

7.   In typical experimental approach, one variable at a time is examined and all other variables are kept constant. In the Taguchi or experimental design approach, all significant variables are mixed and varied in the same experiment. The latter approach permits one to find the important interactions between dependent variables as well as to evaluate the significance of each variable.

8.   The Taguchi approach results in much better understanding and control of the process, particularly the interactions between variables. More importantly, the results point the direction to run complex processes with the minimum variability and explain why some processes go out of control when some parameter is reset.

9.   Without doing the actual experiment, one can only guess as to which variables dominate a process. For baking a cake, the oven temperature and the ingredients (like type of flour) would be dominant along with the pressure (altitude). The cook may be important here also.
        For mowing the lawn, the blade sharpness, blade height versus grass height, blade speed, and blade geometry would probably all be important.
        For washing dishes, the water pressure, the water temperature, the right kind of soap, and perhaps the dish spacing would be most important. Here the operator would not be as important as the design of the machine. The water softness may also be important. The loading of the machine is setup.

10.   Let us assume that you want to drill a 1 cm hole. The drill selected is usually 1 cm in diameter. Undersized holes are not possible until the drill body has worn down, so most of the holes will be oversize. Reground drills often have unequal lip length or rake angles causing the drill forces to be unbalanced, resulting in oversized holes. Assuming you are drilling many holes with many drills, the majority will be oversized, developing a skewed distribution of hole sizes. Do not confuse hole size with hole location. The chisel end of the drill causes problems in obtaining repeated location -- drill "walks" on the surface. Hole location distributions are not necessarily skewed.

11.   There are very few manufacturing processes which receive no inspection during their manufacture. However, some processes that run very reliably and consistently day after day are inspected by the N=2 method. This means that the first and last items are carefully checked, and if these are good, everything in between will be assumed to be good also. Some products, like light bulbs, allow the consumer to be the final inspector. If it works, you keep it. If it doesn't, you take it back to the store and get another.

12.   If the test is destructive (bullets or flash bulbs), if the test is expensive compared to the cost of the item (newspapers), if the item is made in great volumes by reliable or continuous processes (sheets of paper), if the test takes a long time (lifetime test for electronics), then the output is often sampled. Sampling is thus a more economical means to check the quality, but there is always the trade off that, when you sample, you will make errors in judgement about the whole. See questions 13.

13.   You make no error if you say that something has changed and it has, or if you say that nothing has changed and, in truth, nothing has changed. You make a Type I or alpha error when you say that something has changed and nothing has changed (Saying that the process is bad when it is good). You make a Type II or beta error when your sample says nothing has changed and something has changed (Saying process is okay when it is not).

14. You always have some probability of error when you sample (look at a selected few) and then make a decision about all. Both types of errors can be detrimental, even devastating. Sampling inspection systems which miss defects that result in automotive recalls are very expensive and hurt the product's reputation with consumers. Suppose you have a herd of cows. The vet finds a sick cow (sample of one) and condemns all the rest (which are not sick). Or he looks at one cow, finds it well, but the rest have hoof and mouth disease but are not condemned. Either situation is very bad.

15.   It is usually the beta error which leads to legal action since the beta error results in a defective product which was thought to be good, according to the sample. Many sampling systems are designed to protect those who do the inspection against making type I or alpha errors (saying something is bad when it is good) because alpha errors are embarrassing -- stopping the line only to find nothing is wrong. The same is true in general for beta errors -- the system gets blamed for missing the problem, but since the engineer took no action, no blame is directly assigned. However, beta errors can be many times more costly in the long run than alpha errors.

16.   You have a process which is producing many items and you are measuring some characteristic on each item. All the measurements of all the items create a parent population of measurements of individual items. Assuming the distribution of all the items is normal, it has some standard deviation, called σ'. When you take samples from the parent population of size n, you can create distributions of sample statistics. The means of each sample are called $\bar{X}$ and the range of each sample is called R. Thus $\sigma_{\bar{X}}$ and σ_R are the standard deviations of the distributions of the sample means and sample ranges, respectively. These distributions tend to be normal, regardless of the shape of the parent population (Shewhart's Law of Sample Statistics).

17.   This is the process capability index. A value of 0.8 would mean that the process spread (i.e. the variability as measured by the standard deviation) exceeds the tolerance spread (USL - LSL). A value of 1.0 means that these two measures are equal. A value of 1.33 means that the tolerance spread exceeds the natural spread of the process so that all parts being made are within the specification, provided the process is centered.

18.   The bias factor determines how far the mean of the process lies from the intended mean or the minimal. (How good is the aim of the process?)

19.   When the natural variability of the process (6 sigma prime) exceeds the specified total tolerance, you will have a condition which assures that out-of-tolerance products will be made (defectives, scrap, rework, etc.). Has the proper choice of process been made? Can the tolerances be relaxed? Can the process be improved to decrease its variability? (Are there assignable causes of variation which can be eliminated/) Is this a situation where we will have to live with a certain percentage of defective products? Can we automatically sort out the defects? Will a combination of the above kinds of solutions solve the problem?

20.   This factor includes a measure of the process's ability to center itself or to be centered or well-aimed.

21.   Yes, because sample statistics will be normally distributed about their mean.

22.   When a reason for the cause of the variability can be found, one has an assignable cause. A chance cause is inherent to the process and cannot usually be removed, though its effect can be minimized.

23.   It is easier to compute (by hand) and easier to understand, but gives less information about the sample.

24.   The SD of the distribution of sample means, $\sigma_{\bar{x}}$, is equal to the SD of the parent distribution, σ' , divided by the square root of n, the sample size.

$$\sigma_{\bar{x}} = \sigma' / \sqrt{n}$$

25.   Implement the quality control into the process. The idea is to inspect to prevent the defect, not to inspect to find defects after the process is complete. Sorting the bad from the good does not usually remove the cause of the defect.

26.   This is an open-ended design problem for discussion in class.

PROBLEMS FOR CHAPTER 12

1. For demonstration of the question, golf balls have been
selected. The characteristic to be measured is the diameter. A
golf ball is made from hard rubber with a liquid core. It has a
dimpled surface to improve flight accuracy and distance. Its
diameter is specified as 1.68 inches minimum by the Professional
Golf Association. Measurements were made with a 1 to 2 inch
micrometer.

Golf balls are made by a process with a natural total tolerance
of $6\sigma'$ approximately equal to 0.01 inches with an average size of
1.68 inch.

The data given for the golf balls (See chart provided) was for
good used golf balls found on the golf course near one of the
author's home. They were separated by Titleist (TI), Pro Staff
(PS), Top Flight (TF), Pinacle (PI), and Dunlop (DU). Judging
from the sample, the Top Flight ball is either the most popular
or is played by the poorest golfers, as about half of the balls
found were Top Flights. Obviously, this is a mixed sample and
all of the balls were not made by the same machine. However, all
manufacturers use the same basic process.

You can add to the complexity of the process capability study by
using coins and having the student separate the coins by year.
Assuming all the coins have been in circulation, if one measures
weight or thickness at a given point, one should find a wear
factor related to the age of the coin. Thus, the mean should
show a trend -- coins get thinner and lighter with use. But what
about the standard deviation?

2.   This question uses basically the same mathematics as the
last. M&Ms come in different colors as candy-coated chocolate.
The student must decide what to do with samples from a bag having
mixed colors. Ignoring the different colors means that he (or
she) assumes that there is no difference in the process, when in
fact there must be different processes being used to make the
different colors (or different production lines). It is a bit
tricky to measure the thickness or diameter of M&Ms and easier to
measure their weight if the student has access to a scale of
sufficient precision. The difference between this experiment and
the former one is that in the former one, there were only 48
items. Here, there are many items and they are being sampled.
Doing the experiment for weight allows the student to see how the
sample estimate of X' can be used to obtain the estimate of the
true value, which was obtained by weighing all and dividing by
the total number. Questions like "Does thickness have a greater
variability than diameter?" can be addressed by letting some
students measure thickness and some diameter and comparing their
results.

Sample No.	Measurements of Diameter	$\bar{X}$	R
1 TI	1.683 1.675 1.682 1.680	1.6800	.007
2 PS	1.681 1.678 1.681 1.682	1.6805	.003
3 TF	1.676 1.682 1.682 1.679	1.6798	.006
4 TF	1.677 1.680 1.680 1.679	1.679	.003
5 TF	1.677 1.679 1.679 1.678	1.677	.004
6 PI	1.679 1.681 1.682 1.678	1.6800	.004
7 TI	1.678 1.675 1.678 1.677	1.6770	.003
8 TF	1.681 1.680 1.680 1.681	1.6803	.001
9 DU	1.676 1.679 1.680 1.680	1.6800	.004
10 PS	1.677 1.680 1.677 1.677	1.6775	.003
11 TF	1.678 1.677 1.679 1.677	1.6778	.002
12 TF	1.675 1.677 1.678 1.676	1.6767	.004

$$\bar{\bar{X}} = \frac{\Sigma\bar{X}}{12} = \frac{20.1456}{12} = 1.6788$$

$$\bar{R} = \frac{\Sigma R}{12} = \frac{.044}{12} = .00366$$

Estimate of $\bar{X}'$

$$\bar{X} = \bar{\bar{X}} = 1.6788$$

$$1.68$$

Estimate of σ'

$$\sigma' = \frac{\bar{R}}{d_2} = \frac{.00366}{2.059} = 0.0017775$$

$$\Sigma\bar{X} = 20.1456 \quad \Sigma R = .044$$

3.

$$C_p = \frac{USL - LSL}{6\sigma'} = \frac{1.006 - 0.996}{6 \times 0.0021} = \frac{0.010}{0.0126} = 0.79$$

for $\sigma' \cong 0.0021$

$$\beta = \frac{\mu - NOMINAL}{\frac{1}{2}(USL - LSL)} = \frac{0.00085}{0.005} = 0.17$$

$$C_{pk} = \frac{MIN\{|1.006 - 1.002|, |0.996 - 1.002|\}}{3\sigma'}$$

$$= \frac{0.004}{0.0063} = 0.63$$

4.

$$C_p = \frac{USL - LSL}{6\sigma'} = \frac{0.502 - 0.498}{6 \times 0.00067} = \frac{0.00400}{0.00402} = 0.99$$

$$\beta = \frac{\mu - NOMINAL}{\frac{1}{2}(USL - LSL)} = \frac{0.500246 - 0.50000}{0.002} = \frac{.000246}{.002} = .123$$

$$C_{pk} = \frac{MIN\{|USL - \mu|, |LSL - \mu|\}}{3 \times \sigma'}$$

$$= \frac{0.502 - 0.500246}{3 \times 0.00067} = \frac{0.001754}{0.00201} = 0.87$$

5. (See charts provided)

$$\bar{\bar{X}} = \Sigma \bar{X}/25 = 0.716$$
$$\bar{R} = \Sigma R/25 = 0.182$$
$$UCL_{\bar{x}} = \bar{\bar{X}} + A_2\bar{R} = 0.716 + 0.58(0.182) = 0.82$$
$$LCL_{\bar{x}} = \bar{\bar{X}} - A_2\bar{R} = 0.716 - 0.58(0.182) = 0.61$$
$$UCL_R = D_4\bar{R} = 2.11(0.182) = 0.38$$
$$LCL_R = 0$$

Beginning on 6/12, the X chart shows a trend of 8 points below the mean. This suggests that something is wrong and perhaps has been for a while. The R chart shows good control except for point 11 - not due to chance. See Problem 6.

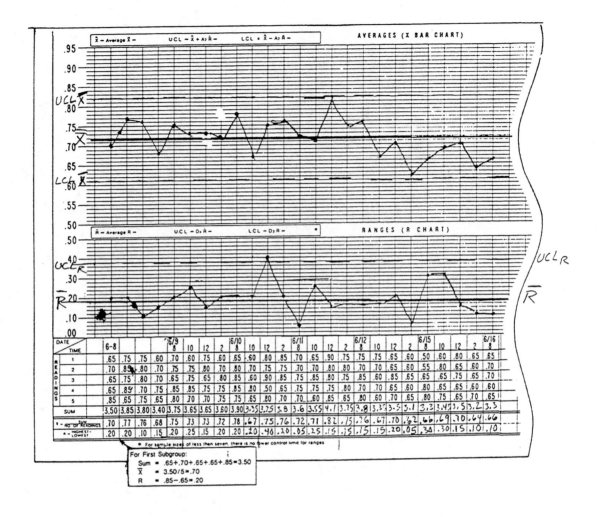

DATE	6-8	6/9				6/10				6/11				6/12				6/15				6/16
TIME	8	8	10	12	2	8	10	12	2	8	10	12	2	8	10	12	2	8	10	12	2	8
READINGS 1	.65	.75	.75	.60	.70	.60	.75	.60	.65	.60	.80	.85	.70	.65	.90	.75	.75	.75	.65	.60	.50	.60 .80 .65 .65
2	.70	.85	.80	.70	.75	.75	.80	.70	.80	.70	.75	.75	.70	.70	.80	.80	.70	.70	.65	.60	.55	.80 .65 .60 .70
3	.65	.75	.80	.70	.65	.75	.65	.80	.85	.60	.90	.85	.75	.85	.80	.75	.85	.60	.85	.65	.65	.65 .75 .65 .70
4	.65	.85	.70	.75	.85	.85	.75	.75	.85	.80	.50	.65	.75	.75	.80	.70	.70	.65	.60	.80	.65	.80 .65 .60 .60
5	.85	.65	.75	.65	.80	.70	.70	.75	.75	.65	.80	.70	.70	.60	.85	.65	.80	.60	.70	.65	.75	.65 .70 .65
SUM	3.50	3.85	3.80	3.40	3.75	3.65	3.65	3.60	3.90	3.35	3.75	3.8	3.6	3.55	4.1	3.75	3.8	3.3	3.5	3.1	3.3	3.47 3.5 3.2 3.3
X̄ = SUM÷NO. OF READINGS	.70	.77	.76	.68	.75	.73	.73	.72	.78	.67	.75	.76	.72	.71	.82	.75	.76	.67	.70	.62	.66	.69 .70 .64 .66
R = HIGHEST − LOWEST	.20	.20	.10	.15	.20	.25	.15	.20	.20	.20	.40	.20	.05	.25	.15	.15	.15	.15	.20	.05	.30	.30 .15 .10 .10

* For sample size of less than seven, there is no lower control limit for ranges

For First Subgroup:
Sum = .65+.70+.65+.65+.85 = 3.50
$\bar{X}$ = 3.50/5 = .70
R = .85−.65 = .20

6.

$\bar{X}'$ is estimated by $\bar{\bar{X}}$ $\bar{X}' \cong 0.716$ (NOMINAL = 0.7)

σ' is estimated by $\bar{R}/d_2$

$\sigma' = \dfrac{0.182}{2.326} = 0.07824$

Note: $\sigma_{\bar{X}} = \sigma'/\sqrt{n} = \dfrac{0.07824}{\sqrt{5}} = 0.03499$

So: $3\sigma_{\bar{X}} = 3 \times 0.03499 = 0.10497 = 0.105$

This value compares well to $A_2\bar{R}$ of 0.1055

$C_p = \dfrac{0.9-0.5}{6 \times \sigma'} = \dfrac{0.4}{0.469} = 0.85$

Ugh!! The process has too much variability and is not well centered.

76

Chapter 13

FUNDAMENTALS OF CASTING

1.   Materials processing is the science and technology by which a material is converted into a useful shape with a structure and properties that are optimized for the intended service environment. More loosely, processing is "all that is done to convert stuff into things".

2.   The four basic families of shape production processes are: (1) Casting, (2) Material removal, (3) Deformation processes, and (4) Consolidation processes. Casting processes can produce extremely complex shapes, but may have defects related to shrinkage and porosity. Material removal processes can have outstanding precision, but generate scrap as the material is cut away. Deformation processes can offer high rates of production, but require powerful equipment and dedicated tools or dies. Consolidation processes can produce large or complex shapes, but the joints may possess properties that are different from the base material.

3.   Cast parts can range in size from a fraction of an inch and a fraction of an ounce to over 30 feet and many tons. Moreover, casting can incorporate complex shapes, hollow sections or internal cavities, and irregular curved surfaces.

4.   In the single-use molding processes, a new mold must be made for each casting. In contrast, multiple-use molds can be used for repeated castings and are generally made of metal or graphite. They are quite costly and their use is generally restricted to large production runs where their cost can be distributed over a large number of castings. For small quantities, the single-use molds would be preferred.

5.   When the molten metal is introduced into the mold, all of the air and gases in the mold prior to pouring and those generated by the action of the hot metal on the mold must be able to escape the mold cavity. This will enable the molten metal to completely fill the mold cavity and produce a fully dense casting that is free from defects.

6.   If the mold provides too much restraint to the solidifying and cooling casting, the casting will crack as it tries to contract while its strength is low.

7.   A casting pattern is an approximate duplicate of the final casting around which the mold material will be packed to form the mold cavity. A flask is the box that contains the molding aggregate. A core is a sand shape that is inserted into the mold to produce internal features in a casting, such as holes or passages. A mold cavity is the void into which the molten metal is poured and solidified to produce the desired casting. A riser is an extra void created in the mold that will be filled with the

molten metal and act as a reservoir to feed metal that will compensate for shrinkage during solidification.

8. The gating system of a mold is made up of a pouring cup, sprue, runners and gates. Its purpose is to deliver the molten metal from the outside of the mold to the mold cavity.

9. A parting line or parting surface is the interface which separates the cope and drag halves of the mold, flask, pattern, or core.

10. The two steps of solidification are nucleation and growth. During nucleation, a stable solid particle forms from the molten metal and forms the beginning of a crystal or grain in the finished casting. During the growth stage, the heat of fusion is continually extracted from the liquid material and the nucleated solid increases in size.

11. At the equilibrium melting temperature, the bulk energies of the liquid and solid states are equivalent. However, for a solid particle to form in the liquid, additional energy must be provided to create the new surfaces or interfaces. Thus, for solid formation to occur generally requires that the temperature drop to several degrees below the melting temperature. Here the change in state from liquid to solid releases sufficient energy that the net result (with the additional energy required to create interfaces) is a movement to a lower energy state.

12. Since each nucleation event produces a grain or crystal in a casting, and fine grain materials possess improved strength and mechanical properties, attempts to promote nucleation would be rewarded by the production of superior castings. This practice of promoting nucleation is known as inoculation or grain refining.

13. In most casting operations, heterogeneous nucleation occurs at existing surfaces, such as mold or container walls, or solid impurity particles within the molten liquid.

14. Directional solidification, in which the solidification interface sweeps continuously through the material, can be used to assure the production of a sound casting. The molten material on the liquid side of the interface flows into the mold to continuously compensate for the shrinkage that occurs as the material changes from liquid to solid.

14. The cooling curve for a pure metal contains information that will reveal the pouring temperature, superheat (the difference between the pouring temperature and the freezing temperature of the metal), the cooling rate, the freezing temperature (thermal arrest), and the solidification times (both total and local).

16.   The term, freezing range, refers to the difference between the liquidus and solidus temperatures, i.e. the temperature range through which the material transforms from all liquid to all solid.

17.   The amount of heat that must be removed from a casting to cause it to solidify is directly proportional to the amount of metal or the volume of the casting. Conversely, the ability to remove heat from a casting is directly related to the amount of exposed surface area through which the heat can be extracted. The total solidification time, therefore can be expressed as proportional to the volume divided by the area to some exponential power - Chvorinov's Rule.

18.   The mold constant, B, in Chvorinov's Rule depends upon the metal being cast, the mold material, mold thickness, and the amount of superheat.

19.   The chill zone of a casting is a narrow band of randomly oriented crystals that forms on the surface of a casting. Rapid nucleation begins here due to the presence of the mold walls and the relatively rapid surface cooling.

20.   The columnar region is clearly the lease desirable. Because of the selective growth process, these crystals are long, thin columns with aligned, parallel, crystal structure. Reflecting this preferred orientation, the properties will be quite anisotropic (varying with different direction).

21.   Dross or slag is the term given to the metal oxides which can be carried with the molten metal during pouring and filling of the mold. Special precautions during melting, pouring and process design can prevent the dross from becoming part of the finished casting. Fluxes can be used to protect the molten metal during melting or vacuum or protective atmospheres can be employed. Dross can be skimmed from the ladles prior to pouring or the metal can be extracted from the bottom of the molten pool. Finally, gating systems can be designed to trap the dross before it enters the mold cavity.

22.   Gas porosity can be eliminated by preventing the gas from initially dissolving in the molten metal, using such techniques as vacuum melting, controlled atmospheres, flux blankets, low superheats, and careful handling and pouring. In addition, dissolved gases can be removed by vacuum degassing, gas flushing, or reaction to produce a removable product compound.

23.   Fluidity is the ability of a molten metal to flow and fill the mold - a measure of its runniness. While there is no single method for its measurement, various "standard molds" can be used where the results are sensitive to metal flow. One approach is to use a long thin spiral that progresses outward from a central pouring sprue. The length of the casting is a direct indication of fluidity.

24. The rate of metal flow through the gating system is important, as is the rate of cooling as it flows. Slow filling and high heat loss can result in misruns and cold shuts. Rapid rates of filling can result in erosion of the gating system and mold cavity and produce entrapped mold material in the final casting.

25. Turbulence of the molten metal in the gating system and mold cavity could promote excessive solution of gases, enhance oxidation of the metal, and accelerate mold erosion.

26. Gating systems can be designed to trap dross and mold material before they enter the mold cavity. Since the lower-density materials will rise to the top of the molten metal, long, flat runners with gates that exit from the lower portion of the runner are effective. Since the first metal to enter the mold is most likely to contain the dross, runner extensions and wells can be used to catch this material and prevent it from entering the mold. Screens or ceramic filters can also be used.

27. Turbulent-sensitive materials, such as aluminum and magnesium, and alloys with low melting points generally employ gating systems that concentrate on eliminating turbulence and trapping dross. Turbulent-insensitive alloys, such as steel, cast iron, and most copper alloys, and alloys with high melting points, generally use short, open gating systems that provide for quick filling of the mold cavity.

28. The three stages of contraction or shrinkage as a liquid is converted into a finished casting are: shrinkage of the liquid, solidification shrinkage as the liquid turns to solid, and solid metal contraction as the solidified material cools to room temperature.

29. Alloys with large freezing ranges have a wide range of temperatures over which the material is in a mushy state. As the cooler regions complete their solidification, it is almost impossible for additional liquid to feed into the shrinkage voids. The resultant structure tends to have small, but numerous shrinkage voids dispersed throughout.

30. A primary concern regarding the contraction of a hot casting after it has solidified is the change in dimensions. In addition, if the product is constrained in a rigid mold, tensile stresses can be induced that may cause cracking. It is best to eject the castings as soon as solidification is complete.

31. By having directional solidification sweeping from the extremities of the mold to the riser, the riser can continuously feed molten metal and will compensate for the shrinkage of the entire mold cavity.

32.   Based on Chvorinov's Rule, a good shape for a riser would
be one with a small area per unit volume. The ideal shape would
be a sphere, but this is rather impractical to the patternmaker
and molder. Therefore, the best practical shape for a casting
riser would probably be a cylinder with height approximately
equal to the diameter.

33.   A top riser is one that sits on top of a casting. A side
riser is located adjacent to the mold cavity, displaced along the
parting line. Open risers are exposed to the atmosphere. Blind
risers are contained entirely within the mold. Live risers
receive the last hot metal that enters the mold. Dead risers
receive metal that has already flowed through the mold cavity.

34.   When using Chvorinov's Rule to calculate the size of a
riser, one makes several assumptions. Since both the riser and
the mold cavity set in the same mold and receive the same metal,
the mold constant, B, is assumed to be the same for both regions.
The equations in the text assume N=2 and a safe difference in
solidification time to be 25%.

35.   Methods developed to assist risers in performing their job
generally have one of two objectives: to promote directional
solidification or to reduce the number and size of the risers.

36.   Casting patterns generally incorporate several types of
modifications or allowances. These include shrinkage allowances
to compensate for thermal contraction, draft to permit pattern
removal, machining allowances, distortion allowance, and
compensation for thermal changes in mold dimensions.

37.   A shrink rule is a simple measuring device that is larger
than a standard rule by the desired shrink allowance. The
measurements on the shrink rule are the final dimensions of the
part after thermal shrinkage has occurred.

38.   A draft or taper is incorporated into casting patterns to
permit the pattern to be withdrawn from the mold without causing
the sand particles to be broken away from the mold surface.

39.   Pattern allowances increase the size of the pattern, and
thus the size and weight of a casting and possibly the amount of
material that must subsequently be removed by machining to form a
finished product. Therefore, efforts are generally made to
reduce the various allowances.

40.   If too large a fillet is used in a casting design, a
localized thick section is produced, resulting in a hot spot in
the casting. These regions cool more slowly than the rest of the
casting and produce localized, abnormal shrinkage, porosity, or
voids.

1.   Plate dimensions are 2" x 4" x 6" ; H/D = 1.5 ; n = 2

t_{riser} = 1.25 $t_{casting}$

$(V/A)^2_{riser}$ = 1.25 $(V/A)^2_{casting}$

$(V/A)_{riser}$ = 1.15 $(V/A)_{casting}$

$$\frac{(\pi/4)D^2H}{2(\pi D^2/4) + \pi DH} = \frac{(1.15)(2 \times 4 \times 6)}{2(2 \times 4) + 2(2 \times 6) + 2(4 \times 6)}$$

Substituting H = 1.5 D into the equation

$$\frac{\pi D^2(1.5D)/4}{\pi D^2/2 + 1.5\pi D} = (1.15)(48/88)$$

$$\frac{3\pi D^3/8}{2\pi D^2} = 1.15 (0.545)$$

3D/16 = 0.627

Therefore:
D= 3.34"   H = 5.02" V_{riser} = 43.98 cu. in.

Yield = Vol. casting / (Vol. casting + Vol. riser)
     = 48 / (48 + 43.98) = 0.52 = 52%

2.   If the riser sits on top of the casting, heat is not lost from either the casting or the riser at their junction. This interface area should be subtracted from both the area of the casting and the area of the riser in Chvorinov's Rule:

$$\frac{\pi D^2H/4}{2(\pi D^2/4) + \pi DH - \pi D^2/4} = \frac{1.15 (48)}{88 - \pi D^2/4}$$

$(3D/14)(88 - \pi D^2/4) = 55.2$

$0.168 D^3 - 18.86 D^2 + 55.2 = 0$

Solving by trial and error, D = 3.25"   H = 4.875
     and V_{riser} = 40.44 cu. in.

Yield = 48 / (48 + 40.44) = .54 = 54%

3.   For the 3" x 5" x 10" solid

$$t_s = B \ (V/A)^2$$

$$11.5 = B \ \frac{(\ 3 \ x \ 5 \ x \ 10)^2}{[2(3x5) \ + \ 2(3x10) \ + \ 2(5x10)]^2}$$

$$11.5 = B \ (150)^2 \ / \ (30 \ + \ 60 \ + \ 100)^2 \ = \ B \ (150/190)^2$$
$$= B \ (0.789)^2 \ \ \ = 0.623 \ B$$

$$B = 11.5 \ / \ 0.623 \ \ = 18.46$$

For a casting of 0.5" x 8" x 8" cast under the same conditions:

$$t_s \ = 18.46 \ x \ (.5x8x8)^2 \ / \ [2(.5x8) \ + \ 2(.5x8) \ + \ 2(8x8)]^2$$

$$= 18.46 \ x \ (32)^2 \ / \ [\ 8 \ + \ 8 \ + \ 128]^2$$

$$= 18.46 \ x \ (32/144)^2 \ \ = 18.46 x (0.222)^2 \ \ \ = 18.46 \ x \ 0.049$$

$$= 0.91 \ minutes$$

CASE STUDY - CHAPTER 13
The Cast Oil-Field Fitting

1.   The binder for the no-bake sand is a polymerizable
alkyd-oil/urethane material. Gases can be evolved from the
binder when it is heated and the polymer material begins to
depolymerize. In fact, there are two possibilities for gas
problems with this material. If the binder had been completely
polymerized during the manufacture of the core, the high
temperature of the cast iron could break down the binder into
small fragments having low molecular weight and low boiling
point, thus producing the bubbles. In addition, this particular
type of binder has a long curing time -- 12 to 24 hours are
required for the polymerization to complete at room temperature.
If the core or the mold were not completely cured, there would
already be low molecular weight, low boiling point, constituents
present that could form gases as soon as the liquid iron entered
the mold cavity.

     The gases are located near a surface, just beneath the
core. It appears that the gas bubbles formed, started to float,
and were trapped by the core.

     Vents could be added to the core and/or mold to give the
gases an easier path to escape through the sand, rather than
becoming trapped in the liquid metal. In addition, we want to
make sure that the binder is completely cured prior to pouring.
A coarser grained sand with a narrow distribution of sand grain
sizes will provide higher permeability and permit easier gas

removal. Finally, a switch to a different type of binder could reduce the amount of gas produced from that of the oil/urethane.

2.   Penetration occurred by liquid metal flowing between the sand grains of the core. It appears that the core was not properly compacted, with relatively large voids between the sand grains. The core may have also had very large sand grains with a very narrow distribution of sizes (although this is contrary to the conclusion of question 1. The core also gets hotter than the mold, since the core is completely surrounded by liquid metal. In addition, the region showing the penetration is adjacent to the gate where it will have received the molten metal first and would have been hotter longer than the remainder of the mold. The long exposure to high heat may have led to the breakdown of the binder and helped the liquid metal penetrate the sand. Finally, the defect was only noted near the bottom of the casting because of the higher metallostatic pressure head (the pressure of the column of molten metal) helping to force the metal between the sand grains.

3.   The enlargement could have occurred because the mold was weak and the high metallostatic pressure crushed the sand, thus enlarging the mold cavity. Better compaction during mold making would produce denser, and stronger, sand. Using a larger amount of binder might also help, but gas problems would tend to become more severe. Another possible cause would be erosion, because the enlargement occurred next to the gate where all of the liquid metal entered the mold cavity. The sand near the gate becomes the hottest, and the binder may have decomposed prematurely. The use of several gates, rather than just one, might help reduce the problem.

4.   Penetration over all of the surfaces is likely due to the sand being too coarse and a narrow distribution of sand grains, or perhaps due to a high pouring temperature. Reducing the pouring temperature would be helpful. Another possible cure would be to use finer sands, perhaps with the addition of silica flour to the aggregate -- although lower permeability and metal-mold defects, such as burn-on might become problems.

5.   Both the molds and the cores could be reclaimed. The binders are organic, and, with luck, most of the organic material will have broken down during the casting and cooling process. If the organic breakdown is not sufficient, some form of reclamation process can be used. A mechanical reclamation system would perhaps fire the sand grains at a hard metal plate, where the impact would break the brittle polymer binder off of the sand grain surface. A thermal reclamation system, in which the sand is heated to a high temperature (usually above $1000^{\circ}F$), will burn off any residual binder. The processed sand is then carefully screened to assure the proper size and distribution of sizes prior to rebonding and reuse.

Chapter 14

EXPENDABLE-MOLD CASTING PROCESSES

1.   A casting pattern is a duplicate of the part to be made, modified in accordance with the requirements of the casting process, metal being cast, and particular molding technique that is being used.

2.   The material used for construction of a casting pattern is determined primarily by the number of castings to be produced, but is also influenced by the size and shape of the casting, the desired dimensional precision, and the molding process. Wood patterns are easy to make and are used when quantities are small. Unfortunately, wood is not very dimensionally stable due to warping and swelling with changes in the humidity. Metal patterns are more expensive, but are more stable and more durable. Hard plastics, such as urethane, have been used, and expanded polystyrene is used for single-use patterns. Expanded polystyrene and wax can be used for single-use patterns.

3.   Split patterns permit the molding of more complex shapes without requiring the hand forming of the parting plane or the use of cutout follow boards. The two halves of the pattern are maintained in proper alignment through the use of tapered pins and holes. Pins in the cope half of the pattern align with holes in the drag half to hold the pieces in proper position.

4.   With a cope-and-drag pattern, the cope and drag halves of the split pattern are mounted onto separate match-plates, thereby permitting larger molds to be handled easier or two separate machines to be simultaneously producing the two portions of the mold.

5.   A loose-piece pattern is frequently used when the object to be cast has protruding sections or geometric features such that a more traditional pattern could not be removed from the molding sand.

6.   The four requirements of a molding sand are: refractoriness, cohesiveness, permeability, and collapsibility. Refractoriness is provided by the basic nature of the sand. Cohesiveness is provided by coating the sand grains with clays that become cohesive when moistened. Permeability is a function of the size of the sand particles, the amount and type of clay or other bonding agent, the moisture content, and the compacting pressure. Collapsibility is sometimes promoted by adding cereals or other organic materials that burn out when exposed to the hot metal to reduce the volume of the solid bulk and decrease the strength of the restraining sand.

7.   The four requirements of a molding sand are not consistent with one another, so good molding sand is always a compromise between the various factors. The size of the sand particles, the amount of bonding agent, the moisture content, and the organic matter are all selected to attain an acceptable compromise. For example, increasing the amount of clay will enhance cohesiveness, but decrease permeability.

8.   A muller is a mixing-type device designed to uniformly coat the grains of sand with the additive agents. The discharge frequently contains some form of aerator which prevents the sand from packing too hard during handling.

9.   Standard tests have been developed to maintain consistent sand quality by evaluating: grain size, moisture content, clay content, compactibility, and mold hardness, permeability and strength.

10.   A "standard rammed specimen" is a 2" in diameter, 2" long sand specimen that is produced by means of a standard and reproducible form of compaction. A sufficient amount of sand is placed in a 2-inch diameter steel tube so that after a 14-pound weight is dropped three times from a height of 2-inches, the final height of the sand specimen is within 1/32 of an inch of 2 inches.

11.   Permeability is a measure of how easily gases can pass through the narrow voids between the sand grains. A casting mold material must possess permeability to permit the escape of air that was in the mold before pouring, plus gases generated from the molding material itself when materials in the molding sand burn, volatilize, or deteriorate when in contact with the hot metal.

12.   The basic size and geometry of the sand grains can be very influential in determining the properties of the molding material. Round sand grains give good permeability and minimize the amount of clay required. Angular sands give better green strength because of the mechanical interlocking of the grains. Large-grain sands provide good permeability and better resistance to high-temperature melting and expansion. Fine-grain sands provide good surface finish on the finished casting. Uniform-size sands give good permeability, while a wide size distribution provides a better surface finish.

13.   When hot metal is poured into a silica sand mold, the silica sand heats up, undergoes one or more phase transformations, and has a large expansion in volume. Since only the surface sand heats up and expands, while the remainder stays cool, the mold experiences nonuniform expansion, and the hot surface may buckle or fold (sand expansion defects).

14.   Penetration of the hot metal between the sand grains can be produced by high pouring temperatures (excess fluidity), high metal pressure (possibly due to excess cope height or pouring from too high an elevation above the mold), or the use of coarse, uniform sand particles.

15.   If sand is placed on top of a pattern and the assembly is then lifted and dropped several times (jolting), the sand is packed firmly around the pattern, with density diminishing as one moves further from the pattern. When squeezing is used, the maximum density is adjacent to the squeeze head. Density then diminishes as the distance from the squeeze head increases. The jolt-squeeze combination combines these two results to produce a more uniform distribution of sand density.

16.   Heavy metal weights are often placed on top of the molds to prevent the sections from becoming separated as the hydrostatic pressure of the molten metal presses upward on the cope. These are used only during pouring and solidification, and can then be removed and placed on other molds.

17.   The vertically-parted flaskless molding machine produces blocks of sand that contain a cope impression on one side and a drag impression on the other. When assembled side-by-side, they produce a complete pattern, with one complete mold being provided per block of sand. Other methods require separate cope and drag segments, thereby requiring two blocks per mold.

18.   Extremely large molds are often constructed in sunken pits, and are often made as an assembly of smaller sections of baked or dried sand.

19.   The two major sources of problems with green sand are low strength and high moisture.

20.   Dry-sand molds lack popularity because of the long times required for drying, the added cost for the drying operation, and the availability of practical alternatives.

21.   In the carbon dioxide-sodium silicate molding method, the carbon dioxide is nontoxic and odorless, and no heating is required for curing. When hardened, however, the sands have poor collapsibility, making shakeout and core removal rather difficult. Here, the heating from the pour actually makes the mold material stronger.

22.   No-bake sands use organic resin binders that cure by chemical reactions that occur at room temperature.

23.   In the shell molding process, a thermoplastic binder is used to bond the sand grains and the "cure" is provided by the exposure of the sand and binder mixture to patterns that have been heated to temperatures in the range of 300-450°F. The heat then cures a layer of mold material adjacent to the pattern.

24. In the V-process, mold strength is obtained through the use of a specially-designed vacuum flask. When a vacuum is drawn, the sand packs to rather high hardness. In contrast, the Eff-set process uses frozen water as a binder, and the molds are poured while in their frozen condition.

25. Cores are used to produce internal cavities or reentrant sections in a casting. These are features that would be extremely difficult to produce by alternative methods.

26. The major problems with green sand cores is their weakness. If they are long or narrow, they are prone to breaking and may not even be able to support themselves.

27. The binder in the core-oil process is a vegetable or synthetic oil. Oven drying causes the oil to cross-link or polymerize, bonding the grains of sand.

28. In the hot-box core-making process, a liquid thermosetting binder and a catalyst are used to bind the sand. When the sand contacts a heated core box, the elevated temperature induces curing within 10 to 30 seconds.

29. Room-temperature curing is the primary attraction of the cold-box process. No-bake and air-set sands have the same advantage, but use a mixed catalyst in place of permeated gas to induce the cure.

30. Shell-molded cores offer excellent permeability since they are generally hollow.

31. Since cores may be nearly surrounded by molten metal, they generally require greater permeability than the base molding sand. All gases must escape through the core prints. Excellent collapsibility is required to permit the core material to be easily shaken out from the interior of the casting.

32. Cores must have adequate permeability and venting to permit the escape of trapped and evolved gases. These properties can be enhanced through the addition of vent holes in the core, or the incorporation of high permeability materials, such as cinders, in the interior of larger cores.

33. Chaplets are small metal devices that provide support for cores and prevent them from shifting in the mold. They should be large enough so that they do not completely melt and permit the core to float, and small enough that the surface melts and fuses with the cast metal.

34.   When plaster molds are made, plaster is mixed with water and hardening occurs by a hydration process. If a high-melting-temperature metal is poured into a plaster mold, the rapid evolution of the hydration water can cause an explosion. Therefore, plaster molds are only used with the lower-melting-temperature nonferrous metals and alloys.

35.   In one of the most popular ceramic molding techniques, the Shaw process, a slurrylike mixture of refractory aggregate, hydrolyzed ethyl silicate, alcohol, and a gelling agent is poured over a reusable pattern. When the mixture has set to a rubbery state, the pattern is stripped, the alcohol is ignited, and the ceramic is fired. The firing makes the mold hard and rigid, but also forms a network of microscopic cracks that provide the desired permeability and collapsibility.

36.   Thin sections can be obtained with plaster and ceramic molds because the mold material retards cooling. High dimensional accuracy, fine detail, the elimination of machining, and exceptionally fine surface finish are all advantages of plaster and ceramic molds.

37.   Graphite molds are often specified for use with highly reactive materials, such as titanium, that would interact unfavorably with many of the more common molding materials.

38.   Rubber-mold casting is limited to the production of small castings of low-melting-point materials, such as wax, plastic, and some low-melting-point metals.

39.   The individual patterns for investment casting are usually made from a molten wax, although plastics and mercury have also been used.

40.   Investment casting molds are preheated prior to pouring to assure that the molten metal will flow more readily to all thin sections and to give better dimensional control by permitting the mold and the metal to shrink together during cooling.

41.   The necessity of removing a pattern from a mold often requires some design modifications, a complex pattern, or special molding procedures. When pattern removal is not required, no draft needs to be incorporated on the pattern, complex single-piece patterns can be used, and it is not necessary to use mold segments, such as a cope and drag.

42.   The full-mold process is attractive for producing small quantities of intricate shaped products, since an expensive pattern is not required and provisions do not have to be made for removal of the pattern from the mold. Castings of almost any size can be made from both ferrous and nonferrous metals. Accuracy may permit the reduction or elimination of finishing operations. Cores and parting lines are eliminated, yields are high, and the backup (support) sand is often directly reusable.

CASE STUDY - CHAPTER 14
Moveable and Fixed Jaw Pieces for a Heavy-Duty Bench Vise

1.   The mechanical properties and size of the piece clearly favor the use of some form of ferrous material, and the size and shape tend to dictate casting. The high elongation can be met by some of the more specialized cast irons, such as ductile cast iron, but also tends to favor the cast steels. Because of the size of the product, some form of expendable-mold sand casting would be likely. While green sand is a possibility the molds will require considerable strength and processes involving stronger molds, such as shell mold will be preferred. Cores will be used to produce the interior channels.

2.   Because of the wide variation in section size, the as-cast products would be expected to have high and complex residual stresses and variability of structure and properties. To relax stresses, achieve uniformity, and attain the desired properties, most of the recommended alloys would require some form of heat treatment. Because the properties are not extremely demanding, a furnace anneal may be all that is required. Normalizing may be used if the variability with section size and location can be tolerated. The replaceable jaws would need the higher hardnesses produced by a quench and temper heat treatment process.

3.   Corrosion resistance to a shop environment is desirable, and would most likely need to be imparted by paint or other form of surface treatment. If paint is selected, both adhesion and durability (including resistance to various oils and solvents) would be selection conditions. A sand blast treatment may be useful in cleaning the surfaces and producing the roughness necessary for enhanced adhesion.

Chapter 15

MULTIPLE-USE-MOLD CASTING PROCESSES

1.   The major disadvantage of the expendable-mold casting processes is the requirement that a separate mold be created for each casting. Variations in mold consistency, mold strength, moisture content, pattern removal, and other factors contribute to dimensional and property variation within a production lot.

2.   Since the multiple-use molds are generally made from metal, the processes are often restricted to the casting of the lower-melting-point nonferrous metals and alloys. Part size is often limited, and the dies or molds can be rather costly.

3.   The reusable molds for permanent mold casting are frequently made from gray cast iron, steel, bronze, or graphite. Aluminum, magnesium, and the copper-based alloys are the metals most frequently cast.

4.   Advantages of the permanent-mold casting process include: a reusable mold, good surface finish and dimensional accuracy, the possibility of controlling solidification to give desired properties, and a fast cooling rate to produce a strong structure. Cores can be used to increase the complexity of the castings.

5.   Permanent-mold castings are generally removed from the mold immediately after solidification because the rigid cavity offers great resistance to shrinkage. Tearing might occur if the part is restrained while cooling.

6.   Permanent molds are not permeable, so venting is usually accomplished through slight openings between the mold halves or the addition of small vent holes. Since gravity is the only means of inducing metal flow, risers are still used to compensate for shrinkage. Both sand and retractable metal cores can be used to increase product complexity.

7.   Slush casting can be used to produce hollow shapes with good surface detail. Wall thickness is variable, so products are largely decorative items, such as candlesticks, lamp bases, and statuary.

8.   Low-pressure permanent-mold casting introduces the molten metal into the die by forcing it upward through a vertical tube. The driving force is a low pressure of 5 to 15 psi applied the molten bath.

9.   Since no risers are used in the low-pressure permanent-mold process (the pressurized feed tube acts as a riser) and the molten metal in the feed tube can be immediately reused, the yields of the process are generally greater than 85%. The metal is exceptionally clean since it is bottom-fed and never passes

through air. Nonturbulent filling further reduces gas porosity
and dross, and directional solidification and pressure feeding
act to minimize shrinkage problems.

10.   Vacuum permanent-mold casting offers all of the advantages
of the low-pressure process (clean metal, low turbulence, low
dross, compensation for shrinkage, high yields, and good
mechanical properties). In addition, the vacuum produces an even
greater cleanliness and low dissolved gas content. Thin-walled
castings can be produced with excellent surface quality.

11.   In low-pressure permanent-mold casting, the feeding
pressures are on the order of 5 to 15 psi. In die casting, the
molten metal is forced into the molds by pressures of thousands
of pounds per square inch and is held under this pressure during
solidification.

12.   Most gravity permanent-mold dies are made from gray cast
iron. This material has great resistance to thermal fatigue and
machines easily. In contrast, die casting dies are generally
made from hardened tool steels, since cast iron cannot withstand
the casting pressures.

13.   Because high pressures might cause turbulence and air
entrapment, lower injection pressures may be preferred, followed
by higher pressure after the mold has filled completely and the
metal has started to solidify.

14.   Hot-chamber die-casting machines cannot be used for the
higher-melting-point metals, such as brass and bronze, and molten
aluminum has a tendency to pick up iron from the casting
equipment. Therefore, the hot-chamber machines are generally
used with zinc-, tin-, and lead-based alloys.

15.   Die casting dies fill with metal so fast that there is
little time for the air in the mold cavity to escape. To
minimize entrapped air problems, such as blow holes, porosity and
misruns, the dies must be properly vented, usually by the
incorporation of wide, thin vents at the parting line.

16.   Since the molten metal is injected into the mold cavity
under pressure and this pressure is maintained throughout
solidification, risers are not incorporated into the mold design
in die casting. Sand cores cannot be used because the pressure
of the molten metal would cause them to disintegrate as the metal
is injected or produce extensive penetration during the cast.
Retractable metal cores can be incorporated into the dies.

17.   By introducing the molten zinc directly into the die cavity
through a heated manifold and heated mininozzles, one can
eliminate the sprues, gates and runners normally incorporated
into a die-casting die, significantly increasing the yield of the
process.

18.   Die casting is characterized by extremely smooth surface finishes, excellent dimensional accuracy, and high production rates. A single set of dies can produce many thousand castings without significant changes in dimension.

19.   In squeeze casting, a precise amount of molten metal is poured into the bottom half of a preheated die set and allowed to partially solidify. An upper die then descends, applying pressure throughout the completion of solidification.

20.   A thixotropic material is a semi-solid (liquid plus solid) material that can be handled mechanically like a solid, but shaped at very low pressures because it flows like a liquid when agitated or squeezed. As an alternative to squeeze casting, it eliminates the need to handle molten metal, and reduces or eliminates many of the molten metal problems, such as gas pickup and shrinkage.

21.   In true centrifugal casting, the metal is forced against the outer walls of the mold with considerable force and solidifies first at the outer surface. Products have a strong, dense exterior surface. Lighter-weight impurities tend to be present on the inner surface, which is frequently removed by a light machining cut.

22.   In centrifuging, centrifugal force is used to force metal from a central pouring reservoir into thin, intricate mold cavities removed from the axis of rotation. Thus, the mold cavities fill under the pressure of centrifugal force. In semicentrifugal casting, a single, axisymmetric casting is poured by introducing metal into the centermost region of a rotating mold. The center pouring region is an integral part of the casting.

23.   In the electromagnetic casting process there is no interaction with a container, and the electromagnetic stirring promotes a homogeneous, fine-grained structure.

24.   Selection of a furnace or melt procedure depends on such factors as: the temperature needed to melt and superheat the metal, the alloy being melted, the desired melt rate and quantity, the desired quality of the metal, the availability and cost of fuels, the variety of metals to be melted, the desire for either batch or continuous melting, the required level of emission control, and the capital and operating costs.

25.   Cupolas are used to produce gray, nodular and white cast irons.

26.   In cupola-melting operations, temperature and chemistry control is somewhat difficult. The nature of the charged materials and the reactions that occur within the cupola can all affect the product chemistry. Moreover, when the final chemistry is determined by the analysis of the tapped product, there is

already a substantial quantity of material within the furnace. Therefore, final chemistry adjustments are often performed in the ladle via ladle metallurgy.

27.   The stirring action, temperature control, and chemistry control of the indirect fuel-fired furnaces are all rather poor. The major attractive feature is the low capital equipment cost.

28.   Arc furnaces offer rapid melting rates, the ability to hold molten metal for any desired period of time to permit alloying (a flux blanket protects the metal), and ease of incorporating pollution control.

29.   Channel-type induction furnaces, where the molten metal circulates through a secondary coil loop and gains heat, offer great ability to provide precise temperature control. These make excellent holding furnaces, where the molten metal must be held for long periods of time at a specified temperature.

30.   The primary functions of a pouring ladle are to maintain the metal at the proper temperature for pouring and to ensure that only quality metal gets into the molds.

31.   Some of the typical cleaning and finishing operations performed on castings include: removal of cores, gates, risers, fins, and rough spots on the surface, cleaning of the surface, and repairing of any defects.

32.   Some types of casting defects can be repaired readily by arc welding. In addition, surface porosity can be filled with a resinous material, such as polyester, by a process known as impregnation.

33.   In metal-casting operations, robots can be used to tend stationary, cyclic equipment, such as die-casting machines. They can be used in finishing rooms to remove sprues, gates and runners, perform grinding and blasting operations, and assist in the heat treatment of castings. They can dry molds, coat cores, vent molds, and clean and lubricate dies. In certain processes, they can be used to dip patterns into mold material slurries and position them to dry. Further implementation might include their use to position the pattern, fill the flask with sand, pour the molten metal, and manipulate a cutting torch to remove the sprue.

34.   Some of the features affecting the cost of a cast product include: the direct cost of material and the energy to melt it, and the indirect cost of patterns, molds, dies, melting and pouring equipment, scrap metal, cleaning, inspection and labor. While the direct costs do not change with the number of castings being produced, the indirect costs must be shared over the production lot. An expensive die may not be justifiable for a small number of castings.

CASE STUDY - CHAPTER 15
Baseplate for a Household Steam Iron

1.   The baseplate must heat to elevated temperatures quickly
and cool to room temperature quickly after use. It must sustain
repeated thermal cycles without deterioration, be light enough to
facilitate ease of use but heavy enough to press out wrinkles
without requiring a pushing force, and provide scratch resistance
to buttons, snaps, rivets and zippers. The material must be
corrosion resistant to steam (and all of the associated
contaminants of various waters), as well as a variety of laundry
products. It must be fracture resistant to dropping from waist
height. The heating element must be thermally-coupled, but
electrically-insulated (NOTE: Normally, thermal conductivity and
electrical conductivity are proportional properties for metals!).
While the key properties are thermal conductivity, corrosion
resistance, and light weight, strength, wear resistance,
toughness, and resistance to creep and thermal fatigue must all
be present at moserate levels. An attractive appearance may be
desired for marketability, and machinability may be required to
produce the necessary holes, threaded recesses, and dimensional
precision.

2.   While aluminum would be the primary material to be
considered, some possible alternatives might be: Stainless steel
(heavy and poorer thermal conductivity), copper alloys (heavier
than steel), and cast iron (heavy). Aluminum alloys offer the
desired low density, low cost, high thermal conductivity, good
corrosion resistance, machinability, and appearance.

      Jumping ahead to marry material and process, the most
attractive process appears to be die casting, and the family of
aluminum die casting alloys is relatively small. Alloy 380
presently accounts for the overwhelming majority of such castings
(probably more than 80%). While other alloys may offer better
thermal conductivity, the enhancement may not be sufficient to
justify deviation from a material that has become the industry
norm.

3.   Because of the integral heating element, production would
most likely be by some form of casting with a "cast-in insert".
Alternative methods include: sand, shell mold, full mold,
permanent mold, plaster mold, investment and die casting, as well
as various methods of forging or powder metallurgy, for which the
insert would have to be a secondary addition.

      Within the realm of reasonable production capability, none
of the processes could produce the completed part to net-shape
directly. Die casting appears to come closest, being capable of
producing the narrow webs and 1/8 inch diameter recesses, but is
not capable of producing the 1/16 inch holes for the steam vents
by means of coring. Surface finish and dimensional precision are
excellent for this application.

4.   The 1/16 inch holes would most likely be added by some form of secondary machining operation, such as drilling. In addition, the larger holes are threaded, but the simplest means of core removal is simple retraction. Therefore, it is likely that these would be made as smooth-bore cavities in the casting, sufficiently undersized to permit the machining of the threads as a secondary operation.

5.   The simple buff and polish may well be the best possible finish. The long-term durability of the teflon coating is questionable, and the anodized layer would produce a darker dull gray finish on the silicon-containing casting alloy. In addition, this layer has been known to flake off due to the differential thermal expansion characteristics and the brittle nature of the oxide material.

Chapter 16

POWDER METALLURGY

1.   Some of the earliest mass-produced powder metallurgy products included coins and medallions, platinum ingots, and tungsten wires. These were followed by carbide cutting tool tips, nonferrous bushings, self-lubricating bearings, and metallic filters.

2.   Automotive applications currently account for nearly 75% of P/M production. Other major markets include: household appliances, recreational equipment, hand tools, hardware items, business machines, industrial motors, and hydraulics.

3.   The powder metallurgy process normally consists of four steps: powder manufacture, mixing or blending, compacting, and sintering.

4.   Some important properties and characteristics of metal powders are: chemistry and purity, particle size, size distribution, particle shape, and the surface texture of the particles.

5.   The most common means of producing metal powders is by melt atomization where molten metal is fragmented into small droplets and the droplets solidify into particles of metal. Any material that can be melted can be atomized and the resulting particles retain the chemistry of the parent material.

6.   Other techniques of powder manufacture include chemical reduction of particulate compounds, electrolytic deposition from solutions or fused salts, pulverization or grinding of brittle materials (comminution), thermal decomposition of hydrides or carbonyls, precipitation from solution, and condensation of metal vapors.

7.   Prealloyed powders can be produced by techniques such as melt atomization, pulverization or grinding, and (in some cases) by electrolytic deposition.

8.   The production of rapidly-solidified material requires immensely high cooling rates and can only be achieved with ultrasmall dimensions. As a result, atomization with rapid cooling and "splat quenching" of a stream of molten metal onto a cool surface are currently the two prominent methods of production. Since much of the "splat cooled" ribbon is then fragmented into powder, powder metallurgy becomes the primary method of producing parts from these materials.

9.   Apparent density is the density of the loose powder to which there has been no application of external pressure. Final density is measured after compaction and sintering and is typically about twice the value of the apparent density.

10.   Green strength refers to the strength of the powder metallurgy material after pressing, but before sintering. Good green strength is required to maintain smooth surfaces, sharp corners, and intricate details during ejection from the compacting die or tooling and subsequent transfer to the sintering operation.

11.   Mixing or blending is performed to combine various grades or sizes of powders or powders of different compositions, or add lubricants or binders to the powder.

12.   The addition of a lubricant improves the flow characteristics and compressibility of the powder at the expense of reduced green strength.

13.   While lubricants such as wax or stearates can be removed by vaporization, the graphite remains to become an integral part of the final product. In the production of steel products, the amount of graphite lubricant is controlled so it will produce the desired carbon content in the final material when it is dissolved in the iron powder.

14.   The goal of the compacting operation is to compress and densify the loose powder into a desired shape. Uniform high density is desired and the product should possess adequate green strength.

15.   Compacting pressures generally range between 3 and 120 tons per square inch, with values between 10 and 30 being most common. The total pressing capacity of compacting presses is the feature that generally restricts the cross-sectional area of P/M parts to several square inches.

16.   During compaction, the powder particles move primarily in the direction of the applied force. The powder does not flow like a liquid, but simply moves in the direction of pressing until an equal and opposing force is generated through either friction between the particles and die surfaces or by resistance from the bottom punch.

17.   When pressing with rigid punches, the maximum density occurs adjacent to the punch and diminishes as one moves away. With increased thickness, it is almost impossible to produce uniform, high density throughout the compact. By using two opposing punches, a more-uniform density can be obtained in thicker pieces.

18.   The final density of a P/M product can be reported as either an absolute density in units of weight per volume, or as a percentage of the theoretical density, where the difference between this number and 100% is the amount of void space still present in the product.

19.   Isostatic compaction is the process in which the powder is
exposed to uniform compacting pressure on all surfaces, and is
usually achieved by encapsulating the powder in a flexible mold
and immersing it in a pressurized gas or liquid. The process is
generally employed on complex shapes that would be difficult to
compact by the faster, more traditional techniques.

20.   In the process of P/M injection molding, the thermoplastic
material is added and heated to provide some degree of fluidity
to the mass of metal powder. This paste-like mixture can then be
injected into a mold cavity to impart the desired shape to a
product.

21.   During sintering, P/M injection molded parts shrink between
15 and 25% as they achieve their final density and properties.

22.   P/M injection molding enables the P/M production of small,
complex components that were previously investment cast or
machined directly from metal stock. Parts can be made with thin
walls and delicate cross sections that would be impossible to
compact in a conventional press.

23.   The three stages of sintering are: (1) the burn-off or
purge -- designed to remove air, volatilize and remove lubricants
and binders, and slowly raise the temperature of the compacts;
(2) the high-temperature sintering stage, with the temperature
being constant; and (3) the controlled cool-down.

24.   Most metals are sintered at temperatures between 70 and 80%
of their melting point. Certain refractory metals may require
temperatures as high as 90% of the melting point.

25.   When sintering, one must slowly raise the temperature of
the compacts in a controlled manner because rapid heating would
produce high internal pressure from heating air entrapped in
closed pores and volatilizing lubricants. This would result in
swelling or fracture of the compacts.

26.   Controlled, protective atmospheres are necessary during
sintering because the fine powder particles have large exposed
surface areas and, at elevated temperatures, rapid oxidation will
occur and impair the properties of the product.

27.   During sintering, metallurgical bonds form between the
particles. In addition, alloys may form, product dimensions will
contract, and density will increase.

28.   Products of HIP techniques generally possess full density
with uniform, isotropic properties that are often superior to
those of products produced by alternative techniques. Near-net
shape production is possible, and reactive materials can be
processed since they are isolated from the environment.

29.   The primary limitations of the HIP process are the cost of "canning" and "decanning" the material, and the long time required for the processing cycle. The sinter-HIP process permits the production of full-density products without the expense and delay of canning and decanning.

30.   Alternative techniques for the production of high-density P/M products include the various high-temperature forming methods, the Ceracon process, and spray forming (also called the Osprey process).

31.   Repressing, coining or sizing operations are generally used to restore dimensional precision. Only a small amount of metal flow takes place.

32.   Repressing cannot be performed with the same tooling that was used for compaction because the compaction tooling is designed to produce an over-sized compact to compensate for the dimensional shrinkage that occurs during the sintering operation.

33.   During repressing, only a small amount of metal flow takes place, and the part retains its starting shape. P/M forging, however, imparts a considerable amount of plastic deformation as the material flows from a simple starting shape to a more-complex shape forging.

34.   While impregnation and infiltration are both processes that fill the permeable void space with another material, infiltration refers to the filling of the voids with another metal, while impregnation employs a liquid, plastic, or resin.

35.   The fracture-related properties, such as toughness, ductility, and fatigue life show the strongest dependence on product density.

36.   When converting the manufacture of a component from die casting to powder metallurgy, it is important to realize that P/M is a special manufacturing process and provision should be made for a number of unique factors. Products that are converted from other manufacturing processes without modification in design rarely perform as well as parts designed specifically for manufacture by powder metallurgy.

37.   The ideal powder metallurgy product has a uniform cross-section, and a single thickness that is small compared to the cross-sectional width or diameter.

38.   In the electrical industry, copper and graphite are frequently combined to provide both conductivity and lubrication. Electrical contacts frequently combine copper or silver with tungsten, nickel or molybdenum, where the material with high melting temperature provides resistance to fusion during the conditions of arcing and subsequent closure.

39.   When compared to cast or wrought products of the same material, conventional P/M products generally possess inferior mechanical properties. This may be an unfair comparison, for if the P/M material and processing is designed to produce a desired product, the desired mechanical properties can often be obtained at lower cost than by alternative techniques. The material, however, is frequently different from that used in wrought or cast equivalents. If full density can be achieved, the properties of P/M products are often superior to their wrought or cast counterparts.

40.   Because of the high pressures and severe abrasion involved in the compacting process, the dies must be made of expensive materials and be relatively massive. The set-up and alignment of punches and dies is frequently a time-consuming process. Production volumes of less than 10,000 identical parts are rarely practical.

41.   The higher cost of the starting material for powder metallurgy is often offset by the absence of scrap formation and the elimination (or reduction) of costly machining operations. Moreover, P/M is usually employed for the production of small parts where the material cost per part is not very great.

PROBLEMS FOR CHAPTER 16

1.   In addition to chemical purity, the key properties or characteristics for material being used in powder metallurgy are those that affect how the powder will, flow, fill space, compact (i.e. respond to pressure), and sinter, as well as those that will directly affect the final properties. These include: surface chemistry, particle size and size distribution, particle shape (and shape distribution), surface texture, and microstructure (or mechanical properties). Since the material is processed as a solid, all of the geometric and property features of the solid become important.

The characterization of starting material for a powder metallurgy process is far more extensive than specifying the starting material for casting (where the material will be melted and both the geometry and the properties will be significantly altered by the process), and forming (where the properties are important, but the starting geometry will be highly altered).

2.   The hot pressing process would not be attractive for the manufacture of conventional P/M parts because loose powder must be protected from oxidation when it is at the elevated temperature. Conventional P/M permits compaction in air because the powder is at room temperature, and reaction rates are acceptably slow. Protective atmospheres are provided during elevated temperature sintering. In hot pressing, some form of "canning" or isolation must be provided and this brings about additional expense and decreased rate of production.

CASE STUDY - CHAPTER 16
Automobile Seat Adjustment Gears

Some of the features that make the gears attractive for powder metallurgy manufacture include: the size (i.e. small), the geometry (flat surfaces with few thicknesses; straight side walls producing segments with constant cross-section; surface finish and precision requirements), high volume mass-production requirements; the ability to produce the gear segments without secondary machining and the associated production of scrap; and the ability to achieve the desired properties.

Alternative processes might include some form of precision casting, finish machining of forged preforms, or machining from sized rods that may have been shaped by extrusion, rolling or cold drawing. More recently, powder metallurgy injection molding may be another alternative.

There are several ferrous powders that are capable of producing the desired properties, however, they are generally alloyed with copper, nickel or other alloy, and require a subsequent heat treatment to achieve the necessary strength and hardness. Several of the iron-nickel-copper alloys possess the necessary properties, along with reasonably good toughness (a desireable property for the proposed application).

Production of the gears would involve mixing, compaction, sintering, and subsequent heat treatment. To avoid reheating, the heat treatment may be integrated into the end of the sintering by decreasing the temperature as the parts approach the end of the sintering operation (to reduce the temperature gradient and related problems) and then directly quenching in oil. A subsequent temper will produce the desired properties. Further surface finishing by techniques such as steam treating to produce a black oxide finish and a final impregnation with oil may enhance appearance and performance.

Chapter 17

THE FUNDAMENTALS OF METAL FORMING

1.   Deformation processes shape metal in the solid state through the rearrangement rather than the removal of material. Unfortunately, large forces are required, and the machinery and tooling can be quite expensive. Large quantities may be necessary to justify the capital expenditure.

2.   Large production quantities are often necessary to justify the use of metal deformation processes because the large forces require costly machinery and tooling.

3.   Independent variables are those aspects of a process over which the engineer has direct control. They are generally selected or specified when setting up the process.

4.   The specification of tool and die geometry is an area of major significance in process design. Since the tooling will produce and control the metal flow, the very success or failure of a process often depends upon good tool geometry.

5.   It is not uncommon for friction to account for more than 50% of the power supplied to a deformation process. Product quality is often related to friction, and changes in lubrication can alter the material flow and resulting material properties. Production rates, tool design, tool wear, and process optimization all depend upon friction and lubrication. In addition, lubricants often act as coolants, thermal barriers, corrosion inhibitors, and parting compounds.

6.   If the speed of a metal forming operation is altered, several changes can occur. Many materials are speed-sensitive and will behave differently at different speeds. Ductility may vary, and many materials appear stronger when deformed at faster speeds. In addition, faster speeds promote lubrication efficiency and reduce the amount of time for heat transfer and cooling.

7.   Dependent variables are aspects of a process determined by the process itself as a consequence of the values selected for the independent variables.

8.   It is important to be able to predict the forces or powers required to perform a specific forming process, for only by having this knowledge can the engineer specify or select the equipment for the process, select appropriate tool or die materials, compare various die designs or deformation methods, and ultimately optimize the process.

9.   The engineering properties of a product can be altered by both the mechanical and thermal history of the material. Therefore, it is important to know and control the temperature of the material throughout the process.

10.   Metal-forming processes are complex systems composed of the material being deformed, the tooling performing the deformation, lubrication at surfaces and interfaces, and various other process parameters. The number of different forming processes is quite large, and various materials often behave differently in the same process. The independent variables interact with one another, so the effects of any change are often quite complex.

11.   The predictive link between independent and dependent variables is generally based on one of three approaches: experience, experiment, or process modeling.

12.   To be truly valid, direct experiments should be full-size at production speeds. Reduced magnitude testing generally alters lubricant performance and thermal effects. Results should be extrapolated to production conditions with caution. In addition, experimentation is costly and time-consuming.

13.   The accuracy of the various process models can be no better than that of the input variables, especially those like strength of material and interfacial friction.

14.   A constitutive equation for an engineering material is an attempt to mathematically characterize the material's behavior under various conditions of temperature, strain, strain rate, and pressure.

15.   Many of the process models describe friction by a single variable of constant magnitude -- i.e. friction is the same at all locations and throughout the entire time of the process.

16.   It is important that the metal-forming engineer know the strength or resistance to deformation of the material at the relevant conditions of temperature, speed of deformation, and amount of prior straining. In addition, he would benefit from information on the formability and fracture characteristics, the effect of temperature and variations in temperature, strain hardening characteristics, recrystallization kinetics, and reactivity with various environments and lubricants.

17.   The value of interfacial friction depends upon such variables as the contact area, speed and temperature. As a result, there is considerable difficulty in scaling down for laboratory experiments or extrapolating up to production conditions.

18.   The friction encountered during metalforming operations is significantly different from that observed in most mechanical operations. In forming, a hard, nondeforming tool interacts with a soft, plastic, workpiece at relatively high pressure. Only a single pass is involved, and the workpiece is often at elevated temperature. Mechanical operations usually involve materials of similar strength, under elastic loading, with a wear-in cycle, and at relatively low temperatures.

19.   According to modern friction theory, frictional resistance is equal to the product of the strength of the weaker material and the actual area of metal to metal contact.

20.   Since the workpiece passes over the tooling only once, wear on the workpiece is generally not objectionable, and may actually be desirable as it produces a shiny, fresh-metal surface. Wear on the tooling, however, alters the dimensions and surface finish of the product and increases the power losses due to friction. Replacement of costly tooling may be required along with lost production during the changeover.

21.   Lubricants should be selected for their ability to reduce friction and suppress tool wear. Other considerations include: ability to act as a thermal barrier, coolant, or corrosion retardant; ease of application and removal; lack of toxicity, odor and flammability; reactivity; thermal stability; stability over a wide range of processing conditions; cost; availability, surface wetting; and the ability to flow or thin and still function.

22.   If one can achieve full-fluid separation between a tool and workpiece, the required deformation forces may reduce by 30 to 40% and tool wear becomes almost nonexistent.

23.   In general, an increase in temperature brings about a decrease in material strength, an increase in ductility, and a decrease in the rate of strain hardening - all effects that would tend to promote ease of deformation.

24.   The temperatures required for hot working generally exceed 0.6 times the melting point of the material on an absolute temperature scale. Cold working generally requires temperatures below 0.3 times the melting point, and warm working is the transition region, between 0.3 and 0.6 times the melting point.

25.   Hot working is deformation under conditions of temperature and strain rate such that recrystallization occurs simultaneously with the deformation.

26.   Hot forming operations do not produce strain hardening and the companion loss of ductility, permitting the material to be deformed by extensive amounts without the likelihood of fracture or the use of excessive force (elevated temperature lowers the strength and deformation does not increase it). In addition,

diffusion is promoted, pores can be reduced or welded shut, and the metallurgical structure can be altered to improve properties.

27.   Disadvantages associated with hot working involve the reactions which may be promoted by elevated temperature, such as rapid oxidation. Tolerances are poorer and the metallurgical structure will be nonuniform if the amount of deformation or thermal history varies throughout the product.

28.   If a metal is deformed sufficiently at temperatures above the recrystallization temperature, the distorted structure (of deformation) is rapidly replaced by new, strain-free grains. The final structure of the metal is that produced by the last recrystallization and the any subsequent thermal history. The production of a fine, randomly-oriented, spherical-shaped grain structure can improve not only the material strength, but also the ductility and toughness.

29.   While the metal grains recrystallize during hot forming, inclusions and nonmetallic impurities do not and serve to impart an oriented or fiber structure (directional properties) to the product.

30.   Heated dies are often used in hot forming operations to reduce the amount of heat loss from the workpiece surface to the tooling and maintain the workpiece temperature as uniform as possible. Nonuniform temperatures may produce surface cracking or nonuniform flow behavior and undesirable properties.

31.   When dies or tooling is heated, the lifetime tends to decrease. Therefore, the upper limit to tooling temperature is generally set by some minimum desired lifetime.

32.   Compared to hot working, cold working requires no heating, produces a better surface finish, and offers superior dimensional control, better reproducibility, improved strength, directional properties, and reduced contamination problems.

33.   Some disadvantages of cold working include: higher forces, required use of heavier and more powerful equipment, less ductility, required surface cleanliness, and the possible need for recrystallization anneals. Detrimental directional properties and undesirable residual stresses may also be produced.

34.   Key tensile test properties that can be used to assess the suitability of a metal for cold forming include: the magnitude of the yield-point stress, the rate of strain hardening, and the amount of ductility that is available.

35.   Springback is an important phenomenon in cold working because the deformation must be carried beyond the desired point by an amount equal to the subsequent springback. Moreover, the amount of springback tends to differ from material to material.

36.   Luders bands or stretcher strains are the ridges and valleys that can form on the surface of sheet metal that has undergone a limited amount of stretching. If the total stretch is less than the yield-point runout, some segments of the metal will undergo deformation and thin by an amount consistent with the entire yield-point runout while other regions resist deformation and remain at the original thickness. Both responses can occur since both require the same applied stress.

37.   Compared to hot forming, warm forming offers reduced energy consumption, less scaling and decarburization, better dimensional control, improved surface finish, less scrap, and longer tool life. Compared to cold forming, it offers reduced forces on tooling and equipment, improved material ductility, and a possible reduction in the number of intermediate anneals.

38.   Isothermal forming may be required for materials, such as titanium, that have strengths that change greatly with a small change in temperature.

PROBLEMS FOR CHAPTER 17

1. a).   Additional costs would include the cost of a heating furnace and the energy costs to achieve the warm working temperature. Tool life would be affected by the combination of increased temperature (decreasing lifetime) and the reduced loads associated with thermal softening (increasing lifetime). Experience would determine which of the above effects would dominate. The reduction in strain hardening could reduce or eliminate the need for intermediate anneals, but consideration should be given toward attaining the desired final properties. Expanded capabilities in terms of size, complexity, and range of possible materials may expand possible markets.

   b).   The conversion from hot forming to warm forming would be accompanied by an obvious savings in energy (heating the workpiece to a lower temperature and heating less material due to higher precision). Additional energy might be saved if it is possible to achieve the desired final properties without requiring a final heat treatment (there is some strain hardening with warm working). Improved dimensional precision and surface finish (reduced scaling and decarburization) can mean savings through a reduction in finish machining and the amount of material converted into scrap. Tool life is increased because of the reduced temperatures and the reduction of thermal shock and thermal fatigue. The forces required for forming will increase by 25 to 60%, so machinery must be more powerful, or the size of products produced on a given machine must be reduced.

2.   Machining operations simply cut through the existing structure, removing the unwanted portion of material. The dimensions of the starting material must be sufficient to contain the crests of the threads. Thread rolling, on the other hand, forms the threads by displacing material from the root of the threads up into the crests. The starting diameter is between that of the root and crest.   The benefits of material conservation and oriented flow continue as discussed in the text, but by cold forming the threads, the effects of strain hardening must also be considered. The deformed material will become stronger, but less ductile. The strengthening can be a significant asset, as long as the accompanying loss of ductility does not make the material too brittle. The residual stress pattern imparted by the deformation can be another concern as it can affect fatigue performance and contribute to failure by stress-corrosion cracking. In addition, it should be noted that the increased strength can be lost if the surface is exposed to elevated temperatures during operations such as hot-dip galvanizing.

CASE STUDY - CHAPTER 17
Repairs to a Damaged Propeller

1).   The specific recommendation would depend upon a number of factors: (1) What is the present condition or structure of the metal? Is it as-cast, age hardened, or annealed? ;(2) What is the ductility of this material in this condition? Can it be mechanically reformed without fracture? ;(3) Would elevated temperature aid in the reshaping? ;(4) Would any subsequent treatment be required after reshaping to restore the desired properties?

     If the material is in the as-cast or annealed condition (one can determine this by hardness testing), and if the material has sufficient ductility, a reshaping may be possible directly. However, one should keep in mind that the propeller has undergone extensive cold working in the initial bending and may not possess sufficient remaining ductility. If insufficient ductility is present, a softening anneal may be required before the reshaping should be attempted. Depending upon the available ductility and the extent of damage, a series of anneal and deform operations may be necessary.

     Finally, consideration should be given to the desired service properties. If the material were directly restored to shape, the bent portions of the propeller would have undergone two rather severe cold forming operations, and would likely be very low in remaining ductility and, therefore, prone to fracture upon any subsequent impact. Since the initial propeller was able to sustain such a severe deformation without fracture, it appears that the initial condition was one with substantial ductility. Therefore, it may be desirable to restore uniform ductility through an anneal after the reshaping.

If the propeller had been age hardened for strength, this heat treatment should be reperformed after the straightening to again produce the strong, homogeneous structure.

Finally, after all heating and cooling, the propeller should be rebalanced to provide smooth running at high RPMs. An out-of-balance propeller can produce excessive loads on bearings and power train components in the engine.

2). In most cases, such a repair can be made, if done properly. The cracked region should be machined out to assure removal of all cracked metal and the exposure of good, clean metal surface. A matching chemistry metal (or near matching to prevent the formation of a galvanic corrosion cell) should then be deposited into the machined groove. Oxyacetylene welding or repair brazing would be the most likely techniques for such a repair.

After deposition, the surface should be rough ground and then fine ground or abraded to produce a smooth surface. Consideration should then be given to the structure of the base metal and the possible effects in the heat-affected zone. If necessary, the entire propeller should be heat treated to produce a uniform structure. Alternately, a stress-relief treatment should be considered to remove potentially damaging residual stresses imparted by the braze. Finally, the propeller should be rebalanced.

If properly performed by an experienced repairman, the repaired propeller will function adequately and will probably cost about half of a new part. Failure to perform a proper repair, however, will result in further cracking problems and the ultimate need to replace the component.

Chapter 18

HOT-WORKING PROCESSES

1.   Metal forming probably began with "tools" as simple as rocks being used to shape bits of naturally-occurring metal. Hand tools and muscle power then gave way to machine processes during the industrial revolution. The machinery further evolved, becoming bigger, faster, and more powerful, and the sources of power also changed. Most recently, computer control and automation have been incorporated.

2.   Various means have been used to classify metal forming process. These include: (1) primary processes that produce intermediate shapes, and secondary processes that produce finished or semifinished products; (2) bulk deformation processes and sheet-forming operations; and hot-working processes and cold-forming processes.

3.   The division of metal forming processes into hot working and cold working is quite artificial. With increased emphasis on energy conservation, the growth of warm working, and new advances in technology, a temperature classification is often arbitrary. Processes normally considered as hot forming processes are often performed cold and cold-forming processes can often be aided by some degree of heating.

4.   At elevated temperatures, metals weaken and become more ductile. With continual recrystallization, massive deformation can take place without exhausting material plasticity. In steels, hot forming involves the deformation of the weaker austenite structure as opposed to the much stronger, room temperature ferrite.

5.   Because the rolls are so massive and costly, and multiple sets of rolls may be required to produce a given product, hot-rolled products are normally available only in standard sizes and shapes for which there is enough demand to permit economical production.

6.   In a rolling operation, friction between the rolls and the workpiece is the propulsion force that drives the material forward. If the friction force is insufficient to deform the material, the material remains stationary and the rolls simply skid over the surface. No deformation is achieved.

7.   Early reductions (with thicker pieces) usually utilize two-high or three-high mills with large diameter rolls. The three-high configuration allows the material to be passed back-and-forth through a single mill without having to stop and reverse the direction of roll rotation. Smaller diameter rolls are more efficient when rolling thinner material, but are less rigid and flex into a distorted configuration. To utilize these more efficient rolls and yet provide rigidity, four-high mills

are used with support being provided by the more-massive backup rolls.

8.   In a continuous or multi-stand rolling mill, it is important that each stand pass the same volume of material in a given time so as to prevent buildup between the stands or tearing of the material being rolled. As the cross-section is reduced, length increases, so the rolls of each successive stand must turn faster than the preceding one by an amount equivalent to the cross-sectional area reduction taken by the previous stand.

9.   Ring rolling is used to produce rings or hoops having a uniform cross-section throughout the circumference.

10.   The rolling of uniform thickness product requires that the gap between the rolls be uniform. Three-point bending occurs when the rolls are loaded in the middle and supported by bearings on either edge. Attempts to compensate by "crowning" the rolls are designed for a specific load, which may vary with changes or fluctuations in material, temperature, lubrication, and other factors. When the thickness is not uniform, the amount of lengthening will not be constant over the entire width, resulting in such defects as wavy edges, wavy center, fractured edges, or fractured center.

11.   Thermomechanical processing consists of simultaneously performing both deformation and controlled thermal processing so as to directly produce the desired levels of strength and toughness in the as-worked product. The heat for the property modification is the same heat used in the forming operation, and the need for subsequent heat treatment is often eliminated. Product properties can be improved and cheaper materials might be employed.

12.   Steam or air hammers use pressure to both raise and propel the hammer. They give higher striking velocities, more control of the striking force, easier automation, and the capability of shaping pieces up to several tons. Computer control can be used to provide specified blows of energy for each step of a process.

13.   Open-die forging does not confine the flow of metal in all directions, so the final shape is dependent upon the manipulation and skill of the equipment operator. Impression-die forging operations confine metal flow in all directions to provide good repeatable control of size and shape.

14.   Open-die forging is not a practical means for the production of large quantities of identical parts because the shape is produced by manipulation and positioning of the workpiece in the hands of a skilled operator (flow of metal is not controlled) rather than by rigid confinement in a set of shaped dies. Each workpiece, therefore, is a separate entity and is not identical to the others.

15.   Because flashless forging involves total confinement of the material within the die cavity, precise workpiece sizing is required along with precise positioning of the workpiece within the cavity and control of the lubrication.

16.   Because counterblow machines permit the excess energy to be dissipated in the form of recoil, there is a reduction in the amount of noise and vibration, two of the major concerns of regulating agencies concerned with the forging industry.

17.   Energy conservation is the primary attraction of processes like the Autoforge and Osprey process. In the Autoforge process, a cast preform is removed from a mold while hot, finish-forged in an impression die, and trimmed to produce a "forged" product. In the Osprey process, atomized droplets are sprayed into a shaped collector to produce a shape that is one forging operation removed from a completed forging.

18.   Dimensions contained entirely within a single die cavity can be maintained with considerable accuracy. Dimensions across the parting plane are dependent upon die wear and the thickness of the final flash. While frequently within several hundredths of an inch, these dimensions are noticeably less precise than dimensions set totally within the die cavity.

19.   Press forging is often preferred to hammer forging when the workpiece is large or thick and the energy of the hammer is insufficient to produce uniform deformation.

20.   Heated dies are usually employed in press forging because the long time of die contact with the hot workpiece would otherwise permit considerable surface cooling and could produce cracking of the surface.

21.   Hammers impart a blow of energy, travel at high speed, and have short time of actual contact. Presses have longer periods of contact and apply a squeezing action or force. Mechanical presses have consistent and reproducible stroke, and are more rapid than hydraulic presses, which are more flexible and can have greater capacity. Since they move in response to fluid pressure, they are controlled by forces or pressures and position is not as reproducible.

22.   Upset forging is the term applied when the diameter of a piece of material is increased by compressing its length.

23.   Upset forging operations are often used to forge heads on bolts and other fasteners and to shape valves, couplings, and a number of other small components, like those illustrated in Figure 18-15.

24.   Automatic hot forging offers numerous advantages. Input material is low cost and production rates are high. Minimum labor is required and scrap production is reduced. The as-forged

structure is often suitable for machining. Tolerances are good, surfaces are clean, and draft angles are low. Tool life is nearly double that of conventional forging. On the negative side, however, is the high initial cost of the equipment and the restriction of large production quantities.

25.   Roll forging is a process by which round or flat bar stock is reduced in thickness and increased in length. A heated bar is placed between two semicylindrical rolls containing shaped grooves, and as the rolls rotate, the bar is squeezed and rolled out toward the operator.

26.   The objective of net-shape or near-net-shape forming is to directly form products that are close enough to specified dimensions that few or no secondary operations are required. Cost savings and increased productivity can be attributed to the reduction in secondary machining operations, reduced quantities of generated scrap, and a decrease in the energy required to produce the product.

27.   The extrusion process offers a number of attractive features. Almost any cross-sectional shape can be extruded, including many that could not be achieved by rolling. Size limitations are few. No draft is required, and the amount of reduction in a single step is limited only by the capacity of the equipment. Frequently only one die is required for a product. Because only a single die change is required to change products, small production quantities are economically feasible. Dimensional tolerances are quite good.

28.   The primary limitation of the extrusion process is that the cross section must be the same for the entire length of the product being extruded.

29.   The area under the direct extrusion curve is proportional to the work required to form the product with billet-chamber friction. The area under the indirect curve is the work required to form the product without frictional resistance. Therefore the "efficiency" of the direct extrusion process could be regarded as the fraction of the total work that is producing deformation. This can be computed as the percentage of the direct curve that is within the indirect region, i.e. the area under the indirect curve, divided by the area under the direct curve, times 100%.

30. In extrusion, the final surface area is considerably greater than the surface area of the starting billet. Therefore, as the material is flowing through the extrusion die, the initial layer of lubricant must spread and thin by a substantial amount, while still functioning as an acceptable lubricant.

31.   In a spider-mandrel extrusion die, the flow of material divides into several channels and then reforms. If the surfaces are fresh, uncontaminated metal, they can be pressed together to form high-quality, virtually undetectable, welds. If a lubricant

were used, the surfaces of the various segments would acquire a coating of lubricant that would prevent the formation of the welds necessary to produce the continuous wall around the hollow shape.

32.   Hot drawing can be used to produce tall, thin cups, by several methods. If the wall thickness can be thinner than the base, drawing with ironing can be employed. If uniform wall thickness is desired, one or more redraws, or multiple-die drawing can be used.

33.   Ironing is the name given to the process where a cup is placed over a punch and driven through a die where the gap between the punch and die is less than the cup material thickness. The cup wall is thinned and elongated, while the bottom thickness remains unchanged.

34.   Steel skelp can be converted into pipe by either butt welding, or lap welding operations. The welding operations occur simultaneously with the hot deformation.

35.   In hot-piercing operations, the billet is forced over a pointed mandrel that is held in place in the roll gap. Since the product must flow over the mandrel and the mandrel must be held rigidly in position, the length of product tubing cannot exceed the length of the mandrel (which is rather limited).

PROBLEMS FOR CHAPTER 18

1.   The assignment here is direct and needs no further explanation.

2. a). While the volume of material remains the same, the surface area changes considerably.

Starting area = circumference times length

$$= \pi \times 5 \text{ inches} \times 12 \text{ inches}$$

$$= 60 \pi \quad \text{square inches}$$

Final area $$= \pi \times 1 \text{ inch} \times 25 \text{ feet} \times 12 \text{ inches/ft}$$

$$= 300 \pi \quad \text{square inches}$$

Final area / Starting area $$= 300 \pi / 60 \pi = 5$$

The final surface area is five times greater than the initial.

b). The cross-section area of a circle is $\pi D^2 / 4$.

The cross-sectional area of a square is L^2
(where L is the length of a side)

If the areas are equal, and D=1, then L will be equal to the square root of $\pi/4$, or $\sqrt{\pi}/2$.

The starting area is still 60π or 188.4 square inches.

The final area, however, is the perimeter times the length, where the length remains at 25 feet.

$$= 4L \times 25 \text{ ft} \times 12 \text{ in/ft} = 2 \times \sqrt{\pi} \times 25 \times 12$$

$$= 1063 \text{ square inches}$$

The area ratio of final to initial is therefore,

Area ratio = 1063 / 188.4 = 5.64, which is greater than for the cylindrical cross-section

c). Since volume is area times length, a reduction ratio of R (change in cross-sectional area) will result in a final product that has area of 1/R and a length that is R times the original. Assume an initial cylinder with diameter and length both equal to 1 unit.

The initial side area is the circumference x length

$$= \pi \times \text{diameter} \times \text{length} = \pi \times 1 \times 1$$

The final side area would be:

$$= \pi \times 1/\sqrt{R} \times R = \pi \sqrt{R}$$

The ratio of the final to initial area would then be:

$$= \pi \sqrt{R} / \pi = \sqrt{R}$$

NOTE: This is in agreement with the answer to part a).

3. a). Force = (.441) x 50,000 x 3.309 = 73,100 pounds

    b). At maximum force of 60,000 pounds, the pressure will be

    $$= 60,000 / \text{Area of the penny}$$

    $$= 60,000 \text{ pounds} / 0.441 \text{ square inches}$$

    $$= 135,900 \text{ psi}$$

c). Since the pressure on the penny, namely 135,900 psi exceeds the yield strength of the thick press plates, the result will be a one-cent copper insert embedded within the surface of the stronger press platen -- NOT A DEFORMED PENNY!

4. Since the thickness of the strip appears in the denominator, the roll separation force will increase with a decrease in strip thickness. If the thickness becomes very small, as with foil materials, the effect can become substantial.

One way to minimize the effect of thin materials is to note that the term is actually proportional to R/t_{av}. Therefore, if the diameter of the rolls can be decreased in proportion to the decrease in thickness, the effect can be canceled.

The various types of rolling mills and their uses, as described in the text, follows this trend. Billets, blooms and thick slabs are rolled on two-high mills with large diameter rolls (often in the range of 22 to 28 inches in diameter). Conventional sheet and strip is rolled on four-high mills with work roll diameters typically in the range of 4 to 10 inches. Foils are rolled on cluster mills with the contact roll being as small as 1/4-inch in diameter, and multiple thicknesses may be rolled simultaneously to increase the total thickness being rolled.

CASE STUDY - CHAPTER 18
Outboard Motor Brackets

1. The requirements for this part include static strength, corrosion resistance to salt and fresh waters, light weight, and resistance to vibration. A variety of engineering materials would be possibilities, including aluminum, titanium, magnesium and even the copper-base alloys. Copper is heavier than steel and would only be recommended if other alternatives failed. The corrosion resistance of magnesium is questionable and the cost and fabrication difficulties do not favor titanium Some form of aluminum alloy would appear to be the most attractive choice -- either a casting alloy or a wrought alloy, depending upon the recommended fabrication process.

2. The geometry 0(size and shape) is such that impression-die forging or some form of casting process would be the obvious alternatives. The various pros and cons can be evaluated with consideration being given to the estimated production quantity. If forging is selected, the recommended material should be some form of wrought aluminum. If cast, the recommended alloy should be selected for compatibility with the process. NOTE: There are aluminum alloys specifically designed for use with processes such as die casting.

3.   If aluminum alloys are used, an age hardening treatment would most likely be required to achieve the desired mechanical strength. This would involve the stages of solution treatment, quenching, and aging.

4.   The corrosion resistance of aluminum would be adequate for fresh water usage, but might be attacked by salt water. Treatment might be utilized to provide aesthetics as well, and in this case, a color anodizing treatment, such as that commonly seen on aluminum softball bats, might be preferred.

Chapter 19

COLD-WORKING PROCESSES

1.   Attractive features of cold forming over hot forming include: no heating is required, surface finish is better, dimensional control is superior, reproducibility is better, strength properties are improved so cheaper materials may be utilized, directional properties can be imparted, and contamination problems are minimized.

2.   Cold-working equipment is usually more powerful than that used for hot-working because the starting material is stronger (no thermal softening), and the material becomes even stronger as it is being formed due to the effects of strain hardening.

3.   Sheet or strip is often given a skin-rolled reduction pass to produce a smooth surface and a uniform thickness, and also to remove the yield-point phenomenon that causes the formation of Luders bands.

4.   The cold rolling of shaped products generally requires a series of shaping operations, each requiring a separate pass through specially-grooved rolls. Since these rolls are usually expensive, two such rolls are required for each pass, and multiple passes are usually required to produce a product, large production quantities are usually required to justify the expense of the shape-rolling process.

5.   If the starting material is a tube, and a shaped mandrel is inserted before swaging, the metal can be collapsed around the mandrel to simultaneously shape and size the interior and exterior of a product.

6.   In a heading operation, the enlarged region is located on the end of the product. In upsetting, the enlarged region is at some location other than an end.

7.   In cold forging, the material is squeezed into a die cavity to impart a desired shape. Since the type of parts produced are those which would otherwise be machined from bars or hot forgings, material waste can be substantially reduced.

8.   With cold forging, production rates are high, dimensional tolerances and surface finish are excellent, and machining can be reduced. Strain hardening can provide additional strength, and favorable grain flow can be imparted.

9.   By combining extrusion and cold heading, the product can be made from a starting stock of intermediate size. Here, the upset head can now be made easily from the starting material size, and the extrusion of the shank reduces the need for extensive machining operations.

10.   In the hydrostatic extrusion process, billet-chamber friction is eliminated, billet-die lubrication is enhanced by the pressure, and in the pressure-to-pressure mode, the pressurized environment suppresses crack initiation and growth and enables the extrusion of relatively brittle materials. Unfortunately, temperatures are limited, sealing problems are common, and complete ejection of the product by the pressurized fluid must be avoided.

11.   Surface friction is the propulsion force in continuous extrusion.

12.   Roll extrusion is typically used to produce thin-walled cylinders with diameters ranging from 3 to 20 inches.

13.   When only one side of a joint is accessible, riveting can be accomplished through the use of either explosive rivets, or pull-type or pop-rivets where the shank on the inaccessible side is expanded mechanically.

14.   One hardened hub can be used to form a number of identical cavities, so only one part needs to be machined to precision. In addition, it is often easier to machine a male shape on the hub as opposed to a female cavity in the die.

15.   During peening, the highly localized blows deform and tend to stretch the metal surface. This surface deformation is resisted by the metal underneath, producing a compressive residual stress in the surface. Since the compressive stresses subtract from applied tensile loads, they serve to impart added fracture resistance to the product.

16.   Burnishing involves rubbing a hard object over the surface of a material under considerable applied pressure. Minute surface protrusions are deformed, producing a smooth, deformed surface.

17.   "Bending" is plastic deformation about a linear axis with little or no change in surface area. When multiple bends are made in a single operation, that operation is often called "forming".   If the axes of deformation are not linear, or are not independent, the process is called "drawing".

18.   When a material is bent, the material on the outside of the bend is elongated, while that on the inside is compressed. Since the material yields first in tension, more deformation occurs by the tensile mode than the compressive one, and the net result is a thinning of the bend.

19.   Springback is the tendency of the metal to unbend somewhat after bending. This is a natural consequence of the outside tension and inside compression of the material and the material seeking to relax these stresses. To form a desired angle, a material must be overbent to compensate for springback.

20.   Press brakes can be used to produce simple bends, complex bends, seaming, embossing, punching, and other operations.

21.   By specifying the largest possible bend radius, the designer can select from a much wider variety of materials for the production of his product and the forming operation will be easier to perform.

22.   Whenever possible, the bend axis should be perpendicular to the direction of previous rolling. If two perpendicular bend axes are involved, the metal should be oriented with the rolling direction at 45° to both axes.

23.   By designing products to have all of the bends with the same bend radius, manufacturers can significantly reduce setup and tooling costs. The same tooling can then be used to produce all bends.

24.   Bottoming dies compress the full area within the tooling, while air-bend dies form the desired geometry through simple. three-point bending. Air-bend tooling is quite flexible since the degree of bend can be varied by a simple change in press position. Bottoming dies, however, produce a more consistent product.

25.   Cold roll forming progressively bends flat strip into complex (but uniform) cross-sectional shapes. Various moldings, channeling, gutters and downspouts, automobile bumpers, and other uniform cross section shapes have been produced. Short lengths of specialized products would be better produced by tools like a press brake, because of the high cost of the roll forming tooling -- multiple sets of profiled rolls.

26.   Rod or sheet can be straightened by two techniques: (1) roll straightening (or roller leveling) which involves a series of reverse bends designed to stress the material beyond its elastic limit, and (2) stretcher leveling in which the material is stretched beyond its elastic limit.

27.   Shearing is the mechanical cutting of materials without the formation of chips or the use of burning or melting.

28.   Sheared or blanked edges are generally not smooth because the cutting tools actually deform the material only to the point where the applied stresses exceed the rupture strength of the remaining material. The remainder of the edge is produced by a metal fracture and has a rough appearance.

29.   If the punch and die (or shearing blades) have proper clearance and alignment and are maintained in good condition, the sheared edges can often be sufficiently smooth to avoid the need for further finishing. Edge condition can be further improved by clamping the stock firmly against the die from above and

restraining the movement of the piece through the die by an opposing plunger or rubber cushion that applies pressure from below the workpiece.

30.   Since fineblanking presses incorporate separate motions and forces for the punch, hold-down or clamping ring, and opposing (or bottom) punch, they are multiple-action machines and are noticeably more complex than presses used in conventional blanking.

31.   Piercing and blanking are both shearing operations in which a curved shearing punch pushes material into a die. They both involve the same cutting action, but when the piece being punched out is the scrap the process is piercing, and when the piece being punched out is the product, the process is one of blanking.

32.   Variations of piercing and blanking that have come to acquire separate names include: lancing, perforating, notching, nibbling, shaving, cutoff, and dinking.

33.   By grinding a slight angle on the face of a piercing or blanking punch, the maximum cutting force can be reduced. Instead of the entire circumference being sheared simultaneously, the angle allows the cut to be made in a progressive fashion, much like the opening of a pull-tab on a beer or soda can.

34.   To produce a uniform cut, it is important that a blanking punch and die be in proper alignment. A uniform clearance should be maintained around the entire periphery.

35.   By mounting punches and dies on independent die sets, they can be positioned and aligned prior to insertion into the press, thereby significantly reducing the amount of production time lost during tool change.

36.   Standard subpress dies can frequently be assembled and combined to produce large parts that would otherwise require large and costly die sets. In addition, when the die set is no longer needed, the components can be removed and used to construct tooling for another product.

37.   A progressive die set consists of two or more punches and dies mounted in tandem. Strip stock is fed through the dies, advancing incrementally from station to station with each cycle of the press performing an operation at each of the stations. Figure 19-50 illustrates a progressive die operation.

38.   In compound dies, a series of operations is performed sequentially at a single station during a single stroke of the press. In progressive dies, the material moves from station-to-station, with separate strokes of the press being required for each operation.

39.　　The term cold drawing can refer to two different operation. For sheet metal, cold drawing involves plastic flow of material over a curved axis, as in the forming of cup-shaped parts. If the stock is wire, rod, or tubing, the term applies to a process where the cross section of the material is reduced by pulling it through a die.

40.　　In tube drawing, rigid tooling is used to accurately size both the inner and outer diameters of the product. In tube sinking, only the outside diameter is directly controlled (there is no mandrel or plug to restrict and size the inner diameter).

41.　　Tube drawing with a floating plug can be used to produce extremely long lengths of tubular product with a controlled inner diameter.

42.　　Straight-pull draw benches are normally employed to produce finite lengths of products that cannot be conveniently bent or coiled. Wire, and smaller products that can be coiled, is generally drawn in a continuous operation on draw blocks where the length of product is limited only by the amount of starting material.

43.　　Because the reduced section of material is subjected to tensile loading in the wire drawing process, the possible reduction is limited by the onset of fracture. In order to affect any significant change in size, multiple draws are usually required.

44.　　Since the metal being deformed by spinning deforms under localized pressure and does not flow across the form block under pressure, the form block can be made of relatively inexpensive material, such as hardwood or even plastic.

45.　　During shear forming, each element of the blank maintains its distance from the axis of rotation. The metal flow is entirely in shear and no radial stretch has to take place to compensate for the circumferential shrinkage. Wall thickness, however, will vary with the angle of that region to the axis of rotation.

46.　　Stretch forming is used to form large sheet metal components that have relatively small production quantities.

47.　　In shrink forming, the metal tends to thicken and, because it is thin, there is a tendency to buckle or wrinkle. In stretch forming, the metal thins and may tear if the stretch is excessive.

48.　　The pressure-ring or hold-down in a deep-drawing operation serves to control the flow of metal and suppress wrinkling, tearing, or undesirable variation in thickness.

49. Because of prior rolling and other metallurgical and process variables, the flow of metal in deep drawing is generally not uniform in all directions. Excess material is often required to assure desired final dimensions, and a trimming operation is generally employed to establish the final dimensions.

50. The Guerin process employs rubber as the female die, providing the pressure necessary to wrap the sheet metal around a male punch. The hydroform process replaces the female die member with a flexible diaphragm backed by hydraulic pressure. Both processes eliminate the female die member to substantially reduce tooling cost.

51. The high energy-release rates needed by the HERF processes can be obtained by: underwater explosions, underwater spark discharge, pneumatic-mechanical means, internal combustion of gaseous mixtures, and the use of rapidly-formed magnetic fields.

52. Two factors account for the low springback observed during high-energy-rate forming. High compressive stresses are set up when the metal is forced against the die, and some elastic deformation of the die occurs under the high applied pressure, allowing the workpiece to become somewhat overdeformed.

53. Common examples of ironed products include brass cartridge cases and the thin-walled beverage containers. Common embossed products include highway signs (like STOP signs) and industrial stair treads.

54. The major limitation to superplastic forming is the low forming rate that is necessary to maintain the superplastic behavior. Typical cycle times may be on the order of 5 to 40 minutes per part. On the positive side, superplastic forming has made possible the economical production of complex-shaped parts in limited production quantities. Deep or complex shapes can be made as one-piece, single-operation pressings, rather than multistep conventional pressings or multipiece assemblies. The required forces are low. tooling is relatively inexpensive, precision is excellent, and fine details can be reproduced.

55. By measuring and evaluating the distorted grid pattern, regions where the area has expanded can be detected as locations of sheet thinning and possible failure. Areas that have contracted have undergone thickening and may be sites of possible buckling or wrinkles.

56. A forming limit diagram is a plot of the major strain and related minor strain on the surface of metal sheet, indicating the conditions for which fracture occurs. Deformation in regions below this line (the forming limit) can be performed without fracture. Deformation which induces strains at or above the line will incur fracture.

57.   Thin complex-shape products can be produced without the use
of metalforming techniques through processes such as electroforming, in which metal is electroplated onto an accurately-shaped
mandrel and stripped free, and plasma spray forming, where molten
metal is sprayed onto a shaped mandrel where it then solidifies.

58.   In general, mechanical drives provide faster action and
more positive displacement control. Once designed, the stroke of
a mechanical press is fixed and cannot be changed. In addition,
the available force varies with position and is greatest near the
bottom of the stroke. Hydraulic drives offer greater forces and
more flexibility of forces, speeds, and strokes. They are
generally slower than mechanical drives and do not offer as great
a control of position or displacement.

59.   Inclinable presses are often tilted to enable ejection of
the finished parts to be accomplished with the aid of gravity or
compressed air jets.

60.   A transfer press is designed to accept a number of die
sets, positioned side-by-side to create a multiple station
(progressive die-type) operation. With each stroke of the press,
each individual station performs its operation on the material
positioned between the dies. The strip material then advances
forward to the next station, where the press undergoes another
cycle. Since all operations are performed simultaneously with
each stroke of the press, one product is made per stroke.

PROBLEMS FOR CHAPTER 19

1.   The wire drawing process can be characterized as continuous
(provided the various segments of incoming product can be butt
welded), but limited in reduction. Since the deformation force
is applied as tension to the reduced product, the maximum
reduction in area (for perfect frictionless conditions) is 62%,
and a typical reduction is between 20 and 50%. The process works
for almost all ductile materials and can be performed at high
speeds. Because of the large surface area and small volume, the
material is rarely heated; most operations are performed at room
temperature. Hydrodynamic lubrication is possible because of the
high relative speed between the workpiece and die.

Conventional extrusion can perform massive reductions in a
single operation (up to a 400 to 1 reduction ratio), but is
limited to finite length segments of starting material (i.e. it
is a piece-rate process). The process can be performed both hot
and cold on both nonferrous and ferrous metals. Because of the
large amount of deformation and the conversion of deformation
energy into heat, the speed of the process is often rather
limited. High strength materials are usually formed hot with
special lubricants to prevent pressure welding to the chamber
and/or die.

Continuous extrusion is an attempt to achieve the best of both worlds -- namely a continuous, high-reduction process. Present techniques are largely limited to the weaker, nonferrous metals, and are usually conducted with room temperature starting stock. Speeds can be relatively fast, provided adequate cooling can be provided to offset the adiabatic heating induced by the large amounts of deformation being performed in a single operation.

CASE STUDY - CHAPTER 19
The Bronze Bolt Mystery

At the onset, the student should recognize a classical ductile, tensile fracture - cup and cone - on the bolts. Secondly, the bolt design violates a rule of upset forging - diameter of the upset μ 1.5 times the bar diameter. In this case, the diameter of the upset area is approximately 2 times the bar diameter.

One might suspect at the onset that the cause of failure is overloading or overtorquing during tightening (installation). However, to determine the cause of the failures, some bolts should be randomly selected from the process and sectioned along the longitudinal axis, metallurgically polished and etched, and then examined under a light or scanning electron microscope. This will reveal the flow patterns in the metal and any cracks in the heads of the bolts. The problem may well be that too large of an upset is being required to form the large bolt head from bar stock of 3/4 diameter. The rules outlined in Chapter 18 have been violated and bronze is not a particularly ductile material.

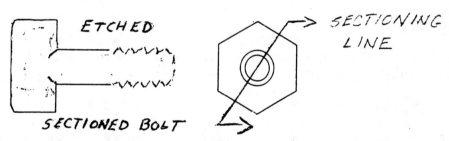

The design should be checked to see if the head must be this big with respect to the shaft. If the head size cannot be reduced, the bolts could be made from larger bar stock and machined down prior to threading. The shaft could also be produced by forward extrusion followed by a cutoff and threading operation. The quantity, however, may be too low for this alternative to be economical. The elimination of the sharp corner where the shaft meets the bolt head would be very helpful in combating this failure problem, but this may require redesigning the pumps that use these bolts, as a radius in this region could prevent the head from seating properly.

Chapter 20

FABRICATION OF PLASTICS, CERAMICS, AND COMPOSITES

1.   Plastics, ceramics and composites are substantially different from metals in both structure and properties. As a result, the processes of fabrication also tend to be different. Many of the fabrication processes can take the raw material to a finished product in a single operation. Large, complex shapes can be formed as a single unit, often eliminating the need for multipart assembly. Joining and fastening operations are quite different from those used on metals. Color, surface finish and precision can often be obtained directly, eliminating the need for surface finishing operations.

2.   Thermoplastic polymers can be heated to temperatures at or near the melting temperature so that the material becomes either a formable solid or a liquid. The polymer can then be cast, injected into a mold, or forced through a die to produce the desired shape. With thermosetting polymers, once the polymerization has occurred, no further deformation can occur. Thus, the polymerization reaction and the shape-forming process must be accomplished simultaneously.

3.   Plastic sheets or plates can be cast between plates of glass. Continuous product can be made by introducing the liquid polymer between moving belts of stainless steel, or into the gap of a rolling mill setup. Tubular products can be made by spinning the liquid against the walls of a rotating mold.

4.   Cast plastics generally contain no filler material, so they present a distinct lustrous appearance.

5.   Blow molding is a process that is used to shape thermoplastic polymers into bottles or other hollow-shape containers. Common thermoplastics for this process include: polyethylene, polyvinyl chloride, polypropylene, and PEEK.

6.   Compression molding is most economical when it is applied to small production runs of parts requiring close tolerances, high impact strength, and low mold shrinkage. Most products have relatively simple shapes because the flow is rather limited. Parts should not contain regions with thick sections because of the long curing times.

7.   Mold temperatures for compression molding typically run between 300 and 400°F, but can go as high as 12000°F. They are generally made of tool steel and are polished or chrome plated to improve material flow and product quality.

8.   Thin sections, excellent detail, and good tolerances and finish are all characteristic of the transfer molding process. In addition, inserts can be incorporated into the product as the liquid resin is introduced at relatively low pressure.

9.   Injection molding is used to produce more thermoplastic products than any other process.

10.   Injection molding of thermoplastic polymer is very similar to the die casting of metal. Sprues and runners channel molten material to the various closed-die cavities where the material solidifies after mold filling. Injection pressures provide rapid filling and prevent premature solidification. The die segments then separate for easy part ejection.

11.   The typical molding cycle in the injection molding of thermoplastics takes from 1 to 30 seconds and is very similar to the die casting of molten metal. Because thermosetting plastics must be held at elevated temperatures and pressures for sufficient time to permit curing, the cycle time for the injection molding of these materials is significantly longer than for the thermoplastics.

12.   By using a hot runner distribution system, the thermoplastic material is kept in a liquid state until it reaches the gate. The material in the runner does not solidify and can be used in the subsequent shot, thereby reducing the amount of scrap and the amount of material that must be heated for each shot (or product). In addition, quality is improved due to the uniformity of temperature and the absence of recycled material in the melt.

13.   In the reaction-injection molding process, two monomers are mixed together as they are injected into the mold. No heating is required, since the chemical reaction between the two components provides the cure and as they polymerize the material actually gives off heat energy that must be removed.

14.   Plastics products with long, uniform cross sections are readily produced by the extrusion process. Common production shapes include solid forms, tubes, pipes, and even coated wires and cables. If the emerging tube is blown up by air pressure, allowed to cool, and then rolled, the product can be a double layer of sheet or film.

15.   Thermoforming is a process designed to shape thermoplastic sheet material into uniform-thickness shaped products, similar to those produced by the embossing of metal sheet.

16.   Rotational molding is used to produce hollow, seamless products of a wide variety of shapes and sizes. These include storage tanks, refuse containers, footballs, helmets, and even boat hulls.

17.   Open-cell foams have interconnected bubbles that permit the permeability of gas or liquid. Closed-cell foams have the property of being gas- or liquid-tight.

18. Rigid-type foamed plastics are used for structural applications, for packaging and shipping containers, as patterns for the full-mold casting process, and for providing rigidity to thin-skinned metal products.

19. Since plastics are poor thermal conductors, little of the heat that results from chip formation will be conducted away through the material or be carried away in the chips. Consequently, the cutting tools run very hot and may fail more rapidly than when cutting metal. In addition, the high temperatures at the point of cutting can cause thermoplastics to soften and swell, possibly binding or clogging the cutting tool.

20. Plastics offer designers a number of unique material properties, including light weight, corrosion resistance, good thermal and electrical insulation, and the possibility of integral color. Some of the design limitations of plastics include: the inability to perform at elevated temperatures of operation, poor dimensional stability, and the deterioration of properties with age.

21. Adequate fillets between the adjacent sections of a mold ensure a smooth flow of plastic into all sections of the mold and also eliminate stress concentrations at sharp interior corners. These fillets also make the mold less expensive to produce and lessen the danger of mold breakage. Rounding exterior edges will act to reduce the possibility of chipping.

22. Uniform wall thickness is desirable in plastic products for several reasons. Since the curing time is determined by the thickest section, it can be optimized for the product if sections are uniform. In addition, nonuniform wall thickness can lead to serious warpage and dimensional control problems.

23. Dimensions parallel to the parting line are contained entirely within a given section of the mold and can possess good dimensional precision. Larger tolerances are required for dimensions which cross the parting line, for they can vary with the fit of the die segments, wear of the dies, and the thickness of any flash that forms.

24. Threaded metal inserts are frequently molded into plastic products because of the difficulty of molding threads in plastic parts and the fact that cut threads tend to chip. The inserts provide strength and allow frequent assembly and disassembly.

25. Since the mold is the reverse of the product, depressed letters or designs would require these features to be raised above the mold. This would require the entire remainder of the mold to be cut away at considerable expense. If the details were raised, only these details would have to be machined.

26.   Dipping can be used to produce relatively thin elastomeric products with uniform wall thickness, such as boots, gloves, and fairings.

27.   Ceramic materials generally fall into two distinct classes, glasses and crystalline ceramics. The glasses are fabricated into useful products by forming a viscous liquid and then cooling it to produce a solid. The crystalline ceramics are fabricated by pressing moist aggregates or powder to a desired shape, followed by drying and bonding by chemical reaction, vitrification, or sintering.

28.   By rapidly cooling the surfaces of hot glass, a residual stress pattern of surface compression can be induced. The resulting glass is stronger and more fracture resistant. Annealing operations can be used to relieve unfavorable residual stresses when they exist, and heating can also be used to promote devitrification - the precipitation of a crystalline phase from within the glass.

29.   Plasticity can be imparted to the crystalline ceramics in a number of ways. Clay products can be blended with water and various additives to permit shaping. Plastic forming involves the blending of the ceramic with additives to make the mixture formable under pressure and heat. The additive material is subsequently removed by controlled heating before the fusion of the remaining ceramic.

30.   Firing or sintering operations provide useful strength to ceramic materials by driving the diffusion processes that are necessary to form bonds between the individual particles. In some cases surface melting or component reactions produce a liquid component that flows to produce a glassy bond.

31.   While machining before firing enables the cutting of weaker material, there is usually a greater concern for the dimensional changes that will occur during firing. Therefore, machining performed before firing is usually rough machining designed to reduce the amount of finish machining to be performed after firing establishes both the final properties and dimensions. In both cases, caution should be exercised relating to the handling of brittle materials that will fail by fracture.

32.   Since the brittle ceramics cannot be joined by fusion welding or deformation bonding, and threaded assemblies should be avoided because of the brittleness of the material, ceramics are usually joined by some form of adhesive-type bonding. Even here, however, the residual stresses can lead to premature failure of the brittle material. As a result, it is best if ceramic products can be produced as single-piece (monolithic) structures.

33.   Many of the manufacturing processes designed to fabricate composites are slow, and some require extensive amounts of labor. There is often a high degree of variability between nominally

"identical" products, and inspection and quality control methods are not well developed.

34. A quality bond can be formed between the distinct layers of a composite material through such processes as: roll bonding, explosive bonding, and the various lamination techniques (adhesive bonding, brazing, etc.).

35. Prepregs are segments of woven fabric produced from the fibers, which is then infiltrated with the matrix material. Subsequent fabrication involves stacking the prepreg layers and subjecting them to heat and pressure to complete the cure of the resin.

36. Sheet-molding compounds are sheets composed of chopped fibers and resin, the sheets being about 0.1 inch in thickness. These can be press-formed in heated dies to provide an alternative to sheet metal where light weight, corrosion resistance and integral color are desired. Bulk-molding compounds are fiber-reinforced thermoset molding materials containing short fibers in random orientation. They are formed into products using processes like compression molding, transfer molding or injection molding.

37. In pultrusion, bundles of continuous reinforcing fibers are drawn through a resin bath and then through a preformer to produce a desired cross-sectional shape. The material is then pulled through a series of heated dies which further shapes the product and cures the resin. Like wire drawing and extrusion, the product is a long length of uniform cross-section.

38. Filament winding is used to produce hollow, container-type shapes with high strength-to-weight ratio. Small quantities of large parts can often be economically made by the filament winding process. The special tooling for a new product (a new form block) is relatively inexpensive. By making these products as a single piece, additional savings can be obtained through reductions in: labor, total manufacturing time, assembly, and tooling costs.

39. Laminated sheets containing woven fibers can be formed to produce the curves required for products, such as boats, automobile panels, and safety helmets. This fabrication is generally performed by processes, such as vacuum-bag molding or pressure-bag molding. Alternative methods include compression molding, resin-transfer molding, and hand layup.

40. When producing products by lamination methods, caution must be taken to avoid the entrapment of air bubbles, and to assure that no impurities are introduced between the layers. The resulting defect or flaw takes the form of incomplete bonding between the layers.

41.   Spray molding utilizes chopped fibers mixed with catalyzed resin. The starting material for sheet stamping is a thermoplastic sheet reinforced with nonwoven fibers. Both chopped and continuous fibers can be used in injection molding using the techniques discussed in Question 42.

42.   Various techniques have been developed to mix reinforcing fibers and resin and permit shaping by injection molding. Chopped or continuous fibers can be placed in the mold cavity, which is then closed and injected with resin. Chopped fibers can be premixed with the resin and simultaneously injected. Continuous-fiber pultruded rods can be sliced into pellets, which then become the subsequent feedstock for the injection.

43.   There are a number of ways to produce fiber-reinforced, metal-matrix composites. Variations of filament winding, extrusion and pultrusion have been developed. Sheet materials can be made by electroplating, plasma spray or vapor deposition onto a fabric or mesh that is then shaped and bonded. Diffusion bonding of foil and fabric sandwiches, roll bonding, and coextrusion are other options. Liquid metal can be cast around fibers through capillary action, pressure casting and vacuum infiltration. Discontinuous fiber products can be made by powder metallurgy techniques or spray forming.

44.   In ceramic matrix composites, the matrix is a brittle material and failure occurs by fracture. A primary purpose of the "reinforcement" is often to impart toughness rather than strength. One means of imparting toughness is to prevent (or interrupt) the propagation of carack across the matrix. By designing weak interfaces, propagating cracks are diverted along the interface, rather than crossing it and continuing their propagation.

45.   When fiber-reinforced materials must be joined, the major concern is the lack of continuity of the fibers in the joint area. The thermoplastic resins can be welded. Thermoset materials require the use of mechanical joints or adhesives.

PROBLEMS FOR CHAPTER 20

1.   There are a number of possibilities here inclusing the aligned fibers in the shafts of golf clubs (pultruded or extruded), the laminates of woven sheets in skiis, the various sheet-type products in racing car bodies, and the continuous fiber reinforcements included in both the frame and handle regions of tennis rackets. Advances are continually being made in both the materials and processes, and sporting goods is one of the most active and competitive markets for the employment of composite materials.

CASE STUDY - CHAPTER 20
Fabrication of Lavatory Wash Basins

1). A cast iron basin would be easy to form and could be made in a variety of intricate shapes. Internal passages for the overflow could be made an integral part of the product through the use of cores. The product would be relatively cheap and could be made available in a range of colors through the use of porcelain enameling. Detriments include the undesirable weight of the product (viewed from both shipping and installation), the inherent brittleness of the material, and the need for extensive surface preparation and finishing.

A deep drawn steel basin would be attractive for a number of reasons, including its lightweight and availability in a variety of colors. The overflow drain would require additional operations to shape and join the related components. Consideration would have to be placed on the joining method so as to prevent the introduction of objectionable surface marks. Resistance seam welding would be a likely candidate. Corrosion prevention would require a protective surface finish, such as porcelain enamel.

The stainless steel basin would be very similar to the steel basin just described, except that no surface finishing would be required. Corrosion resistance and appearance would be excellent, and the product could be polished to a reflective finish. The forming (drawing) of stainless is a bit more difficult than the forming of low-carbon steel, and the resulting product would be restricted to the range of metallic finishes, including bright and satin.

The ceramic basin has long been a standard lavatory product, but has some associated limitations. Compared to the sheet metal approaches, the production rates are rather low. The wall thickness is large, providing some degree of weight and bulk, but this approach provides the added benefit that the basin is frequently a stand-alone product, requiring only a wall or floor mount (instead of a supporting stand or cabinet). Since the basic material is porous, a sealing coating is required, and offers the opportunity to provide a variety of colors. The material is somewhat brittle, and will crack or fracture upon large impacts.

While polymeric materials were originally substandard from the viewpoint of scratch and abrasion resistance and durability, considerable improvement has been made in recent years. Products can now be made that resemble natural marble and offer durability, and even the opportunity to remove damage and refinish the product through mild abrasive and polishing techniques. Wall thickness must be large to provide the rigidity and durability, but the material is sufficiently light weight that this is not a significant detriment. The material is a good thermal insulator, is available in a range of integral colors, offers excellent surface finish, and requires no additional coatings.

2).   The cast iron basin would most likely be sand cast with
sand cores providing the passageways for the overflow and drain.
Steel and stainless steel would probably be formed by successive
deep draw operations. The drain and overflow holes would be
punched and the overflow passage would require the connection of
tubing or the seam welding of a second formed section onto the
original piece. A ceramic basin would probably be formed by slip
casting, and the drain and overflow would be incorporated into
the mold. A polymeric basin could be shaped by either casting or
molding. Again, it would be attractive to incorporate the drain
and overflow into the original design and manufacturing process.

     Each of the above methods, with the possible exception of
cast iron, have features that would make them attractive in the
marketplace.

     The simple low carbon sheet metal basin is inexpensive,
available in a range of decorator colors, and is easily integrated in the countertops and washstands frequently used in most
lavatories. The porcelain enamel coating is durable and will
resist most stains and potential damage. Its lightweight would
be attractive from the viewpoint of shipping costs, and it could
be packaged in a variety of inexpensive means, such as a simple
cardboard box.

     The stainless steel sink appeals to only a limited portion
of the market. For this reason, we might wish to consider it
primarily as an addition or expansion to another manufacturing
system.

     Ceramic whiteware is well established in the marketplace
and can be designed to be either stand-alone or incorporated into
a counter or stand. The full variety of colors is available
through the porcelain enamel finish. However, the rate of
production may be lower than with the alternative sheet metal
methods.

     Finally, the polymeric materials are capable of offering
several unique features, such as their ability to simulate
marble. There is no distinct surface coating, so the color and
properties exist throughout. They can also be manufactured as a
larger unit in which the basin is an integral component of a
continuous countertop. Scratch and abrasion resistance continue
to be an area of some compromise.

3).   A porcelain enamel coating (or similar surface protection)
would be required for the cast iron and steel sheet metal basins,
as well as the ceramic product where it seals the porosity and
provides an attractive finish. The stainless steel material
forms its own natural protective oxide, and the polymeric materials can usually be designed to provide the desired surface
properties directly.

The stainless steel basins would be available in only a limited range of finishes and could not be produced in a variety of colors. Many polymeric materials cannot offer the desired levels of scratch and abrasion resistance, and stain resistance may also be a problem. With the continued development of polymers, however, many of these problems have been sufficiently overcome.

The porcelain enameling of cast iron generally requires special processing, since the carbon in the material reacts with oxygen at the firing temperature to produce gases which cause bubbles and blisters in the surface. With low carbon sheet steels, this becomes less of a problem.

4.)   Actual selection of one of the above solutions would require an assessment of the company's expertise, available equipment, the specific nature of the targeted marketplace, and the desired level of production. Deep drawn steel, ceramic whiteware, and polymeric materials would all be attractive.

Chapter 21

FUNDAMENTALS OF CHIP-TYPE MACHINING PROCESSES

1.   This is a localized plastic deformation process that is not completely bounded and involves very large strains at very large strain rates. It cannot be generalized since the nature of the fundamental shear process is highly material dependent. The behavior is further complicated by the variable nature of the inputs and the large number of input variables. The boundaries that do exist (tool-chip interface and work material) are difficult to characterize. Finally, it is difficult to obtain reliable and consistent measurements from experimental setups and even more difficult to extrapolate these results to shop floor production machines.

2.   One must consider cutting speed, feed, and depth of cut as input parameters and their relationship to things like CT and MRR. The cutting tool design, the workholding devices and the machine tool's design will all influence the dynamics of the process. The nature of the operation must also be considered. The material being machined is most important, as is the cutting tool material itself. The cutting lubricant may also be important as some machining processes cannot be accomplished without a cutting fluid.

3.   Turning, facing, boring, shaping, planing, milling with a fly cutter, some modes of deep hole drilling, and other variations of lathe operations (cutoff or recessing for example) are single point tool operations. Plunge or form turning is a single point tool operation. The rest are multiple edge, and this includes drilling, milling, broaching, sawing, filing, and the many forms of abrasive machining.

4.   The RPM of the workpiece, N_s, and the feed rate of the tool, f_r, determine the feed (in inches per minute) of the tool with respect to the work. $N_s f_r = f$. In turning and milling, the feed is related to the speed. Speed and feed are dependent in turning because increasing the speed means increasing the RPM which increases the rate the tool moves past the work. The feed per revolution is not changed. In milling, changing the cutting speed (increasing the cutter RPM) automatically decreases the feed per tooth of the cutter if the table feed (in/min) is not changed. In broaching, the feed is built into the tool, and will be the same regardless of the speed. In sawing, filing, and grinding, it is not directly related, but rather indirectly. For example, in sawing the feed is often controlled by gravity, pushing the blade into the work as the teeth cut and carry away chips. For the same cutting speed, you can slow up the feed by lightening the load. For shaping, the speed and feed are independently set.

5.   The operator needs to know the feed rate of the table (in/min) as this is an input to the machine. The other feed is the feed per tooth and it is selected depending upon the specific type of milling, the cutting speed, and the specific tool material. The feed per tooth is the maximum amount of material each tooth can remove. The table feed, $f_m = f_t \, n \, N$ where f_t is the feed per tooth, n is the number of teeth in a cutter (teeth/rev) and N is the RPM as determined from $N = 12 \, V /(ÐD)$ for a cutter of diameter D.

6.   A shear-front lamella structure is developed by very narrow shear fronts which segment the chip material into very narrow lamellae. The mechanism is developed out of the compression deformation which precedes the shear. If the material is already cold worked, very little additional compression deformation is needed to activate the shear process. If the material is annealed (or as-cast), the compression deformation is extensive, causing the workpiece to bulge and upset prior to shearing. The shear fronts have micron-spaced periodicity and are the result of many dislocations moving at the same time. The onset of shear begins at the shear plane (defined by ϕ) and moves at the angle ψ to form the chip. This process is microscopic and not visible to the naked eye, except in very special circumstances. The primary dislocation mechanism appears to be one of dislocation pileups against the cell structure produced by compression deformation or prior work hardening of the workpiece material.

7.   The metalcutting process has been labeled an adiabatic shear instability, meaning that the heat input and heat dissipation are balanced, or that there is excess heat which results in softening (lowering the strength of the material) so that the shear instability can take place. However, the metalcutting observed in Figure 21-12 is taking place at such low speeds, such a mechanism appears to be unlikely. At faster cutting speeds, adiabatic shear may be responsible for the large saw-tooth structures seen in chips as the elastic energy is rapidly dissipated over a shear front.

8.   Oblique cutting is what is typically done in almost every machining process, with the exception of experimental setups designed to eliminate one cutting force, thus converting oblique (3 force) cutting to orthogonal (2 force) cutting. The exceptions to this is industrial practice are broaching and slab milling with a straight tooth cutter. Orthogonal machining can be converted to the oblique machining simply by canting the cutting edge with respect to the direction of motion of the tool.

9.   The exact equation for turning is:
$$MRR = 12 \, V \, f_r \, (D_1{}^2 - D_2{}^2) / 4 \, D_1$$
The approximate equation for turning is:
$$MRR = 12 \, V \, f_r \, d$$
and it assumes that the depth of cut, d, is small compared to the uncut diameter of the workpiece, D_1. See page 656.

10.   The mechanics can become quite complicated when a radius is used rather than an edge. Almost all of the analysis work in metalcutting assumes a zero radius cutting edge.

11.   The magnitude of the strain and strain rates are very large for metal cutting compared to tensile testing. Metal cutting strain is on the order of 1 to 2 compared to tensile testing's 0.20 to 0.40 and metal cutting strain rates are 10^5 to 10^9 in/in/sec compared to tensile testing's 10^{-2}.

12.   Titanium is very strain rate sensitive. The faster you deform it, the stronger it behaves. This causes machining problems because of metal cutting's large strain rates.

13.   Cast iron has a structure that is filled with flake graphite. These flakes produce regions which act like sharp-cornered flaws or voids which concentrate the compression stresses. The shear fronts cannot cross these regions. Under the large strains, the metal fractures through the flake and the chips come out segmented or in fractured chunks.

14.   In metal cutting, shear stress is a material constant. This means that it is not sensitive to changes in cutting parameters or cutting process variations. Once this value is known for a metal, it can be used in basic engineering calculations for machining statics (forces and deflections) and dynamics (vibrations and chatter).

15.   The primary or largest force is always the cutting force, F_c which is in the direction of the cutting speed vector, V.

16.   The energy, F_c V is divided into shear (actually compression and shear) to form the chips (about 75%) and secondary shear and sliding friction at the tool/chip interface.

17.   The energy that produces plastic deformation does so through the production of dislocations, which multiply and move. The energy in the dislocations is returned to the metal as heat when the dislocations absorb each other (annihilate). In short, the energy is converted into heat.

18.   It can be estimated from:
    a). the specific horsepower (See equation 21-26)
    b). the shear stress (Assume $F_c = 2 F_t$ and solve
            equation 21-20 for F_c)
    c). the specific energy (Solve equation 21-29 for F_c)

19.   The rate of wear (on the flank or the rake face) of the tool is most directly influenced by cutting speed. The higher the cutting speed, the shorter the life of the tool. This is because increasing V directly drives up the temperature, and increasing the temperature of a tool can rapidly increase wear rates. See Section 22.4.

20. As the temperature goes up, the hardness (resistance to penetration) decreases. See Figure 21-22.

PROBLEMS FOR CHAPTER 21

1. $\tan \phi = \dfrac{r_c \cos \alpha}{1 - r_c \sin \alpha}$; $= \tan^{-1} r_c$ for $\alpha = 0$

$$\widehat{\tau_s} = \frac{F_s}{A_s} = \frac{(F_c \cos\phi - F_t \sin\phi)}{t\omega / \sin\phi}$$

$$\mu = \frac{F}{N} = \frac{F_t}{F_c}$$

$$HP_s = \frac{F_c}{396,000 \, f_r \, d} = \frac{F_c}{396,000 \, t\omega} \quad \text{for tube cutting}$$

Value of $\widehat{\tau_s}$ from Figure in Text $\approx 75,000$

Value of HP_s from Table in text is 0.67

SUMMARY OF RESULTS

Run #	ϕ	F_s	A_s x1000	$\widehat{\tau_s}$	μ	HP_s
1	18.3	220.4	0.978	70,760	0.89	0.85
2	20.9	187.8	0.978	68,502	0.90	0.80
3	23.1	247.6	1.47	66,083	0.80	0.70
4	23.1	286.3	1.47	76,412	0.80	0.72
5	24.6	318.0	1.962	67,470	0.69	0.67
6	24.1	328.6	1.962	69,187	0.73	0.70

2. ρ = density of metal in the chip in lbs/ cu. in.

wt = weight of a chip fragment of length, l_c

Assume $\dfrac{\text{wt of chip}}{l_c \times w_c \times t_c} = \rho$

Since l_c is measured and w_c is assumed to be w,

$$t_c = \frac{wt}{l_c \; w_c \; \rho}$$

Then $\quad r_c = t \,/\, t_c$

3. Calculated for Run #5

$F_c = 510, \quad f_r = 9.81, \quad d = 0.200, \quad \phi = 24.6$

$r_c = 0.458, \quad F = F_t = 350, \quad F_s = 318$

$U = \dfrac{F_c}{f_r d} = \dfrac{510}{0.00981 \times 0.200} = 259{,}938 \text{ psi}$

$U_f = \dfrac{F \, r_c}{f_r d} = \dfrac{350 \times 0.458}{0.00981 \times 0.200} = 81{,}702 \text{ psi}$

$U_s = \dfrac{F_s V_s}{f_r d V} = \dfrac{F_s \, \cos \alpha}{f_r d \, \cos(\phi - \alpha)} \quad\quad \text{but} \quad = 0$

$\quad = \dfrac{F_s}{f_r d \, \cos \phi} = \dfrac{318}{0.00981 \times 0.200 \times 0.9} = 180{,}008 \text{ psi}$

4.

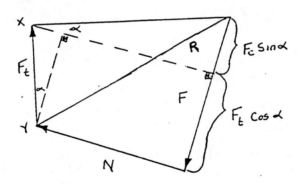

$N = F_c \; \cos\alpha - F_t \; \sin\alpha \quad$ from the diagram
$F = F_c \; \sin\alpha + F_t \; \cos\alpha \quad$ from the diagram

5.

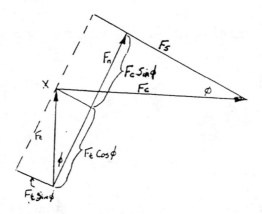

$$F_s = F_c \cos\phi - F_t \sin\phi \quad \text{from the diagram}$$
$$F_n = F_c \sin\phi + F_t \cos\phi \quad \text{from the diagram}$$

6.

$$\varepsilon = \frac{2 \cos\alpha}{1 + \sin\alpha} = 2 \quad \text{for } \alpha = 0$$

$$\frac{1}{r_c} = \frac{1}{.458} = 2.18$$

So these values are quite close to each other

7. From Figure 21-18, the shear stress for Inconel 600 is at 105,000 psi.

$$\tau_s = \frac{F_c \sin\phi \cos\phi - F_t \sin^2\phi}{tw} \quad , \quad w = 0.250 \text{ in.}$$

So $t = 0.020$ in and $t_c = 0.080$ in.
Then $r_c = 0.02/0.08 = 0.25$

$$\text{Tan } \phi = \frac{r_c \cos\alpha}{1 - r_c \sin\alpha}$$

α would be known to the engineer, but here a good assumption would be $\alpha = 0$, since Inconel is very hard to machine and a carbide tool is likely to be used with zero back and side rakes.

So Tan $\phi = 0.25$, $\phi \cong 14$ degrees, is a good estimate

$$105,000 = \frac{F_c (.24)(.97) - F_c/2 (.0576)}{(0.02)(0.2)}$$

$$F_c \cong 2100 \text{ lbs.}$$

If F_c is estimated from HP_s (using 1.8 for HP_s),
$F_c = 2851$

8. For rough machining, the:
   Cutting speed ranges from 200 sfpm to 800 sfpm
   Feed rates from 0.010 ipr to 0.085 ipr
   Depth of cut from 0.125 in to 0.675 in

   MRR_{min} = 12 Vfd (Equation 21-3)
   = 12 (200) (0.01) (0.125) = 3 in^3/min

   MRR_{max} = 12 (800) (0.085) (0.675) = 550 in^3/min
   (NOTE: This would require a machine tool with a very
   large motor capable of delivering several hundred
   horsepower.)

   For finishing:

   MRR_{min} = (12) (700) (0.005) (0.0125) = 0.525 in^3/min

   MRR_{max} = (12) (1600) (0.015) (0.0675) = 19.44 in^3/min

9. The MRR of 550 in^3/min is very large.

   MRR = 12 V f_r d, so 550 = (12) V (0.005) (0.675)

   V = 13,580 ft/min, which is very large.

   This is probably aluminum being machined.
   Let's try HP_s = 0.25.

   $$HP = \frac{F_cV}{33,000} = \frac{(10,000)\ (13,500)}{(33,000)} = 4115\ hp$$

   This value is out of sight! -- which should indicate to
   the engineer that something is amiss here. All of the values
   -- HP, F_c, and V are at least an order of magnitude too
   large!

10. HP = 24 hp   MRR = 550 in^3/min

    HP_s = 24/550 = 0.0436

    Published values for 1020 steel range from 1.1 to 3.4, with
    1.4 being typical for 1020 steel at 200 DHN. This estimate
    is way out of the expected range. Values of HP_s this small
    have not been experimentally observed.

In terms of labor dollars saved, 35 minutes per piece times $30.00 per hour yields a savings of $700 for 40 pieces, which certainly covers the cost of the cutter ($450). However, the machine on which the job has been done may not be capable of performing at the increased speeds needed for the carbide tooling. The cutting speed can probably be increased from 130 sfpm to 300 sfpm with carbide. TiN coated HSS tool inserts or a TiN coated, solid HSS cutter should be given serious consideration. If a change in cutters is being considered, getting a larger cutter with more teeth will decrease the cutting time, assuming the feed per tooth remains the same. That is:

$$CT = (L + A) / (n N f_t) = (L + A) \quad D / (n \, 12 \, V \, f_t)$$

Increasing the cutting speed will increase the power needed to cut the 4140. Does the machine have sufficient HP_m ?

AISI 4140 is usually a relatively hard material and if interrupted cutting is involved, be prepared for some early tool failures due to impact fracture. Some test cuts will have to be made to insure that the surface finish is satisfactory.

Check to see how this part is going to be used in service. If it is a critical part in a fatigue environment which has been performing satisfactorily in the past, be aware of the possibility that switching to carbide could alter the service performance as the new process will undoubtably put a different residual stress distribution into the surface (See Chapter 40). Also check the tool-life performance of the HSS operation versus that of the proposed carbide to get a line on tool changing and tool replacement costs. Hugh is wrong when he says that the tool probably will not be used again, since most carbide cutters sold today have replaceable insert tool bits and can be readily retooled and reused long after the HSS tool has been worn out and discarded.

In summary, either Joseph or Hugh could be right, but there is more to changing cutting tool materials than economics.

Chapter 22

CUTTING TOOLS FOR MACHINING

1.   The most important material property for cutting tools is hardness. The tool must be harder than the material being machined to prevent rapid wearing and early failures.

2.   Hot hardness is the ability to sustain hardness at elevated temperatures. See Figure 21-22.

3.   Impact strength is a material property which reflects the ability of a material to resist sudden impact loads without failure. It is a combination of strength and ductility and is measured by the energy absorbing capability of the material. The two tests used for impact testing are the Charpy and the Izod Impact test. The general term for impact strength is toughness.

4.   Many cutting tools experience impacts during routine cutting processes. Interrupted cuts are common in milling. Cutting tools may also impact on hard spots or hard surfaces of a material.

5.   HIP is hot isostatic pressing, a powder metallurgy process used to make cutting tools, particularly carbides. See Chap. 16.

6.   Primary considerations in tool selection include: What material is going to be machined, what process is going to be used, what are the cutting speeds, feeds, and depths of cut needed, what is the tool material, and what are lubricants going to be used. See Figure 22 - 2 for complete answer.

7.   A hard, thin, wear-resistant coating is placed on a tough, strong, tool material. Such composites have good impact strength and good wear resistance.

8.   Cermets are a relatively new cutting tool material compared to composed of ceramic materials in a metal binder. See Figure 22-9 for a comparison of cermets to other tool materials.

9.   CBN is manufactured by the same process used to make diamonds. The powder is used as a coating for carbide blanks in the same way poly-crystalline diamonds are made. CBN powder is sintered and compacted onto a carbide substrate, diced with a laser into segments and the segments brazed into pockets in a standard tungsten carbide insert. The CBN layer is about 0.020 inches thick.

10.   F. W. Taylor developed the experiments which lead to the Taylor tool life equation, developed the principles of scientific management and stop watch time study, developed the tool grinder methodology for grinding specific angles on cutting tools, and is considered to be one of the founders of Industrial Engineering. He was also the first United States tennis doubles champion, dispelling the myth he had bad eyesight.

11.   Cast cobalt alloy tools would be made by investment casting, due to the high temperatures of the alloys.

12.   The compacted powders are compressed into a solid of uniformly fine grains. If cobalt is used as a binder, the solid cobalt dissolves some tungsten carbide, then melts and fills the voids between the carbide grains. This step is called sintering.

13.   When the cobalt powders melt and fill the voids between the carbide grains, they "cement" the carbide grains together. This is an old term still used in the cutting tool industry to describe sintered, powder metallurgy tools.

14.   The ground inserts are more precise - have less variability from tool insert to tool insert -- so that there is very little difference between tools. This is important when changing tools in automatic equipment or rotating the insert in an indexing tool holder. Therefore, the tool does not have to be reset when the insert tip is changed. Pressed inserts may vary in size as much as .005 inches and may carry this size change into the process.

15.   The chip groove is placed on the rake face directly behind cutting edge. Depending upon the depth of cut, the chip groove can make the land in front of the groove act as a controlled contact surface and modify the cutting process. It can cause the shear angle to increase and therefore reduce the power and cutting forces. It can also cause the chips to bend sharply and fracture into short segments which makes chip disposal easier. See Figures 22-5, 22-6, and discussion on page 625.

16.   As shown in Figure 22-12, a groove forms at the outer edge of the cut during the machining of materials with a hard surface or a surface with hard particles in it. The groove is called the depth of cut line or the DCL since it forms at a distance from the cutting edge equal to the depth of cut.

17.   a) High speed steel will deflect the most - smallest E
       b) Ceramic will resist penetration the most - hardest
       c) High speed steel is the most ductile
       d) Carbides are the strongest in compression.

18.   Tools get hot and expand during machining. Different materials have different coefficients of expansion. The layers are graded with respect to thermal coefficients of expansion to reduce the probability of thermal cracking of the coats. Some layers are also used to promote bonding between the materials.

19.   For high speed steel, black oxide and nitriding are quite common but TiN of HSS is becoming very popular. See Figure 22-8. Coating carbides with TiN and TiC and other materials is popular now using CVD. Aluminum oxide coating is becoming more popular. Ceramics are usually not coated or surface treated.

20.   The reaction forms hydrogen chloride which can affect the impact strength and other material properties.

21.   Lowering the coefficient of friction at the tool/chip interface reduces the secondary deformation. This has the effect of reducing the friction force, F. The reduction of F results in rotation of the force circle (or an increase in the shear angle). For a given cutting geometry, this means a reduction in F_S (because $F_S = \widehat{\tau_S} A_S = \tau_S$ t w/ Sin ϕ) and F_C (because F_C is a function of F_S). Thus, the tools run at lower forces and lower temperatures and last longer.

22.   Diamond reacts chemically with ferrous materials while CBN does not.

23.   The coefficient of variation is the ratio of the standard deviation of a statistic distribution to the mean of the distribution. A large value indicates that the process which produced the data for the distribution has a large amount of process variability.

24.   Tool life varies from tool to tool even when the tools are being used under identical conditions. Lifetime is a random variable, whether we are talking about tools, people, tires, or light bulbs. The random variable nature of tool life means that predicting tool death will be very difficult.

25.   Metal cutting tool life data tends to be log-normal and have a coefficient of variation of .3 to .4. Compare this to values for yield strength data or UTS data which have values of .03 to .05.

26.   Machinability is defined many different ways. The two most common ways are: <u>machining specific horsepower</u> (HP_S) which reflects the power needed to remove a cubic inch of metal per minute - the more difficult metals will have higher numbers for specific horsepower; and <u>machinability numbers based on tool life comparisons</u>. A material is selected as the standard. A material which can be machined faster with the same tool life as the standard material has a higher rating than the standard. So the first measure is based on equal volume of material removed and ignores tool life, and the other on equal tool lives, ignoring power consumed. Other measures of machinability have been proposed using ease of chip removal and surface quality as criteria.
       In the 1970's, one of the authors (Black) tried (without much success) to get people to think about flow stress, τ_S, as a machinability standard (like UTS) and developed a prototype machine to determine flow stress values for various materials. Many of the values given in Figure 21-18 came from this research.

27.   From the earliest measurements of F. W. Taylor, it has been known that cutting fluids provide a cooling action for the tool. Because of the nature of the process, few usable cutting fluids provide a lubrication action to the tool/chip interface but this

has not deterred us from calling them "lubricants". Cooling
helps the tool maintain its hardness and thereby reduces the rate
of tool wear. The cutting fluid also acts as the means to void
the process of chips, protects the newly machined surface from
corrosion and cools the workpiece (reducing expansion and
distortion) and the chips (hot chips are dangerous). Additives
can be placed in cutting fluids which alter the chemistry of the
workpiece surfaces and thereby influence the dislocation behavior
(alter free surface conditions) and the shear process itself.

PROBLEMS FOR CHAPTER 22

1.   The constants can be found by selecting two points from the
plot and equating VT^n to VT^n with n being the unknown. After n
has been found, solve for C. $VT^{0.14} = 46.7$. The tool material
was high speed steel, because n = 0.14, which is a typical value
for steel.

2.

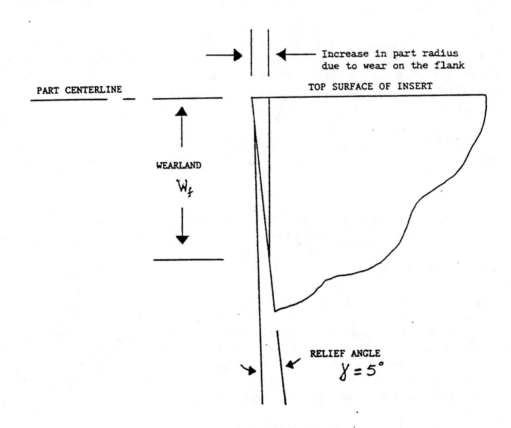

INSERT FLANK WEARLAND, W_f	WORKPIECE DIAMETER GROWTH (5° RELIEF)	$= \tan 5° \times W_f \times 2$
.010	.002	
.020	.004	
.030	.005	
.040	.007	

DEFLECTION OF PART NOT INCLUDED.

FACING OPERATIONS, GROWTH ONLY 50%.

146

3. a). A = side rake angle
       B = side relief angle
       C = end relief angle
       D = back relief angle
       E = nose radius of the tool
       F = side cutting edge angle
       G = end cutting edge angle

   b). The back rake angle is negative

   c). This is probably a brazed insert tool - carbide - for
       turning

4.

Tool Life Values Machining AA390 Aluminum Alloys

Workpiece Material	Cutting Tool Material	C	n	V_{20} m/min	V_{30} m/min	V_{60} m/min
Sand Cast	COMPAX Blank	1906	0.32	731	642	514
Permanent Mold	COMPAX Blank	1588	0.33	591	517	411
Permanent Mold**	COMPAX Blank	1211	0.23	608	554	472
Sand Cast	Tungsten Carbide K-20	329	0.21	175	161	139

**Note: With flood applied cutting fluid on Heidenreich and Harbeck lathe.

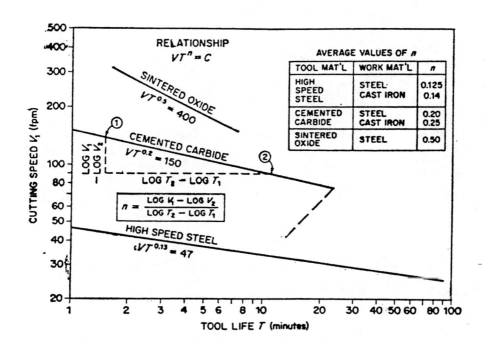

NOTE: The n values for the new COMPAX cutting tools are
      similar to what has been observed for carbides and
      oxides.

147

5.   Tube turning is orthogonal machining when the side cutting edge angle is 0°. Orthogonal cutting becomes oblique or three force cutting when the tool is given a lead angle or a SCEA. The SCEA or lead angle is computed by

$$\text{Cos}\,\theta = .250/.289 \quad \text{Therefore,} \quad \theta = 30 \text{ degrees.}$$

The underformed chip thickness, t, is

$$t = \text{feed rate} \times \text{Cos}\,\theta$$

The results are shown below for four feed rates and four SCEAs.

The effect of increasing the SCEA is to thin the chip, dissipate the heat, protect the insert nose radius, reduce insert notching at the DCL. However, the increased radial forces can cause chatter.

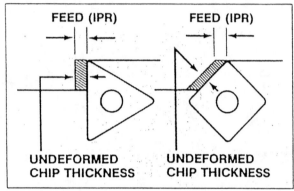

Effect of lead angle on undeformed chip thickness.

UNDEFORMED CHIP THICKNESS AT VARIOUS LEAD ANGLES AND FEED RATES

FEED RATE (IPR)	LEAD ANGLE			
	0°	15°	30°	45°
.005	.005	.005	.004	.004
.010	.010	.010	.009	.007
.020	.020	.019	.017	.014
.030	.030	.029	.026	.021

6. a).

	SiN	PCBN	Comments
Tips needed per part	12	12	Operation needs 12 tips for each cylinder block (6 bores/block)
Tool life parts/tool	200	4700	PCBN has a 23.5:1 tool life advantage
Cost/tip	$1.25	$28.50	SiN has a 22.8:1 cost advantage per tip
Cost/part for tooling	$0.075	$0.0727	PCBN has a $0.0023 advantage per part
Tool coat per year	$23,400	$22,682	This is the tooling cost (just for the inserts) for one station of a transfer line. See page 1137 for discussion.

Would you make the change based on an annual savings of $718? There will be additional savings from tool insert changing time. Probably you should not change unless you can find other advantages, such as surface finish (quality) or less downtime on the transfer line to change inserts.

b). The basis for change can often be based on the cost per part for tooling, but such changes are not usually made without careful examination of the products over the life of the tool. Because the bores in the cylinder block are such a critical, high precision element in the part, changes are fully evaluated before recommendations to change are made. In this case, cylinder blocks would have to be completely assembled and tested (including life testing) before a tooling change like this would be made. This is why cutting tool salesmen are usually engineers. Figure 41-5 on page 1127 shows a transfer for engine blocks. How would you like to be the cutting tool salesman for this machine?

CASE STUDY - CHAPTER 22
Indexable Tool Insert

The cutting speed has such a great influence on the tool life compared to the feed or the depth of cut that it greatly influences the overall economics of the machining process. From the table, we observe that the Al_2O_3 LFG tool is cutting at over three times the speed of the uncoated carbide inserts. For a given combination of work material and tool material, a 50% increase in speed results in a 90% decrease in tool life, while a 50% increase in feed results in a 60% decrease in tool life. A 50% increase in depth of cut produces only a 15% decrease in tool life. So, in limited horsepower situations, depth of cut and then feed should be maximized while speed is held constant and horsepower consumed is maintained within limits. As cutting speed is increased, the machining time decreases but the tools wear out faster and must be changed more often. In terms of costs, the situation is as shown in Figure 22-18, which shows the effect of cutting speed on the cost per piece.

The total cost per operation is comprised of four individual costs. Machining costs, tool costs, tool changing costs, and the handling costs. The machining cost is observed to decrease with increased cutting speed because the cutting time decreases. Cutting time is proportional to the machining costs. Both the tool costs and the tool changing costs increase with increases in cutting speeds. The handling costs are independent of cutting speeds. Adding up each of the individual costs results in a total unit cost curve which is observed to go through a minimum point. For a turning operation, the total cost per piece, TC, equals:

$$TC = C_1 + C_2 + C_3 + C_4$$
$$= \text{Machining cost + tool cost + tool changing cost + handling cost}$$

Expressing each of these cost terms as a function of cutting velocity will permit the summation of all of the costs.

$$C_1 = CT \times C_o$$ where C_o = operating cost (\$/min) and
CT = cutting time (min/piece)

$$C_2 = (C_t \times CT) / T$$ where T = tool life (min) and
C_t = initial cost of the tool (\$)

$$C_3 = (T_c \times C_o \times CT)/T$$ where T_c = time to change tool (min)

C_4 = labor, overhead, and machine tool costs consumed while part is being loaded, unloaded, tools advanced, machine broken down, etc.

Since $CT = L/N f_r$ for turning

$$= \pi DL / 12 V f_r$$

and $T = (C/V)^{1/n}$ rewriting Equation 22-2

TC can be expressed in terms of V:

$$TC = \frac{L \pi DC_o}{12 \, VF_r} + \frac{C_t V^{1/n}}{t \, C^{1/n}} + \frac{t_c C_o V^{1/n}}{t \, C^{1/n}} + C_4$$

To find the minimum, take $d(TC)/dV = 0$ and solve for V.

$$V_m = \left[\frac{1}{n} - 1 \right] \left[\frac{C_T + (C_o \times t_c)}{C_o} \right]$$

Table CS-22 shows the results of Brian's analysis of the four different tools, all used for turning hot-rolled 8620 steel with triangular inserts. Brian assumed that the length of cut was about 24 inches. This table should be carefully studied so that each line is understood. Note that the cutting tool cost per piece was three times higher for the low force groove tool over the carbide, but really of no consequence since the major cost per piece comes from two sources -- the machining cost per piece and the nonproductive cost per piece.

Brian was correct in his belief that the coated tools could save some money. Now, the next question is, what can Brian do to reduce the total cost per piece? Begin the discussion with Figure 41-4.

TABLE CS.22 -	Cost Comparison of Four Tool Materials, Based on Equal Tool Life			
	Uncoated	Tic-coated	Al_2O_3-coated	Al_2O_3 LFG
Cutting speed (surface ft/min)	400	640	1100	1320
Feed (in/rev)	0.020	0.022	0.024	0.028
Cutting edges available per insert	6	6	6	3
Cost of an insert ($/insert)	4.80	5.52	6.72	6.72
Tool life (pieces/cutting edge)	192	108	60	40
Tool change time per piece (min)	0.075	0.075	0.075	0.075
Nonproductive cost per piece ($/pc)	0.50	0.50	0.50	0.50
Machining time per piece (min/pc)	4.8	2.7	1.50	1.00
Machining cost per piece ($/pc)	4.80	2.7	1.50	1.00
Tool change cost per piece ($/pc)	0.08	0.08	0.08	0.08
Cutting tool cost per piece ($/pc)	0.02	0.02	0.03	0.06
Total cost per piece ($)	5.40	3.30	2.11	1.64
Production rate (pieces/hr)	11	18	29	38
Improvement in productivity based on pieces/hr (%)	0	64	164	245

Source: Data from T. E. Hale et al., "High Productivity Approaches to Metal Removal," *Materials Technology*, Spring 1980, p. 25.

Chapter 23

TURNING, BORING AND RELATED PROCESSES

1.   In turning, the work rotates and the tool is fed parallel to the axis of rotation.

2.   Cylindrical, conical, contoured, tapered, and knurled surfaces can be produced externally. Internal turning is called boring.

3.   In form turning, the shape of the tool defines the shape of the surface and the tool is usually fed perpendicular (or plunged) to the axis of rotation. See Figure 23-18 for examples of form turning.

4.   Facing employs a tool that is wider than the desired cut width and the workpiece is not separated into two parts by the process as is the case in a cutoff operation. In both cutoff and facing, the tool feeds perpendicular to the axis of rotation.

5.   Knurling, a common lathe operation, usually does not make chips - it cold forms the pattern into the surface.

6.   It is not possible to provide the proper rake angles on all portions of a complex form tool. Typically, small or even zero back angle tools are used, so the cutting forces will be large. In addition, small increases in depth of cut result in very large force increases because the cutting volume is large (long cutting edge in contact); so depth of cut must be set (and held) small to prevent large deflections and chatter during machining.

7.   Swing is the maximum diameter of a workpiece that can be rotated in a lathe.

8.   A hollow spindle permits long bar stock to be fed through it into the workholding device (chuck or collet) much more quickly than if individual piece parts are used. The drawbar for the collet also must pass through the spindle. See Figures 23-4 and 23-38.

9.   The carriage supports the toolholder and provides the feed motion to the tool.

10.   Feed is specified in inches (or millimeters) per revolution and is a direct input into the lathe.

11.   Either the feed rod or the lead screw drives the carriage. The feed rod system usually has a friction clutch in which slippage can occur. The lead screw provides positive ratios between carriage movement and spindle rotation and no slippage is allowed. The lead screw is for thread turning.

12.   Work in a lathe can be held between centers, held on mandrels which are then held between centers, held in 3 or 4 jaw chucks, or held in collets. Workpieces can also be directly mounted on a face plate attached to the spindle and are occasionally mounted on the carriage and even in the tailstock assembly (very rarely).

13.   A device called a dog is attached to the work and the tail of the dog fits into a hole in the face plate, which is directly attached to the spindle. See Figure 23-39.

14.   After the work has been turned, the surfaces will be tapered rather than cylindrical. See Figure 23-8.

15.   Hot rolled stock usually has an oxide scale on it, which is rough. Clamping on nonround, rough surfaces like this can damage the collet jaws and destroy the accuracy of the collet.

16.   A steady rest is mounted on the ways and is stationary while a follow rest is mounted on the carriage and thus translates with it as it carries the tool. Both are commonly used on long, cylindrical workpieces.

17.   A four-jaw independent chuck can be adjusted to clamp work of almost any shape within its capacity. Such a chuck requires more time to adjust than a three-jaw chuck, and it is not self-centering.

18.   Minimizing the overhang of the tool improves the rigidity of the setup and reduces the tendency of the tool to deflect which in turn reduces the tendency to chatter and vibrate. Remember, deflection is a function of length of overhang cubed in cantilever beams, so a small change in overhang length can greatly change deflection and vibration tendencies.

19.   The workpiece tends to climb over the tool, sharply increasing the depth of cut, rapidly driving up the cutting forces, and ultimately stalls out the machine, often with dangerous or costly breakage of the lathe or the tool or the workpiece.

20.   Tapers may be turned by: (1) use of the compound rest, (2) set-over of the tailstock, (3) use of a taper attachment.

21.   The material removal rate is a function of speed x feed x depth of cut. Assuming speed is kept constant, heavier depths of cut or heavier feeds will reduce the number of cuts or passes that have to be made, which reduces the number of adjustments or resettings of the tool which have to be made. Increasing either feed or depth of cut will increase the cutting force. Increasing depth of cut will have less effect on tool life than increasing the feed. Either way, the total machining time is usually less. In addition, the surface finish is usually better with the lighter feed.

22.   The rpm of a facing cut is based on the largest diameter of the workpiece, utilizing the correct (selected) cutting speed.

23.   See response to question number 18 above. Boring tools have large overhangs and are thus more subject to deflection, vibration, and chatter problems. Reducing the feed (or the depth of cut) reduces the cutting forces and thereby the deflection problems.

24.   As the cutting rate increases, the surface finish usually improves - See Figure 23-9. However, large nose radius tools tend to chatter more.

25.   The BUE will cause the tool to make heavier depth of cuts than expected, so the part may come out undersize.

26.   In tooling a multiple spindle screw machine, it is important to have the machining operations at each spindle require the same amount of time. As shown in Figure 23-18, this time balancing can be very difficult to do. There will be one operation having a processing time larger than any of the other operations. The 2nd position drill or the 4th position tap are probably the operations with the longest operation times (about 5-7 seconds).

27.   Two objectives of boring are: to enlarge a hole to a desired size or to assure that the resulting hole is concentric with the axis of rotation.

28.   In drilling, the drill can drift or shift off center. This is due to the chisel end of the drill not cutting and the drill being deflected as it starts the cut. In boring, the hole is the result of the rotation of the workpiece about its axis, which remains fixed.

29.   On vertical boring machines, the weight of the workpiece is down on, and supported by, the table; whereas, in a lathe, it must be supported and rotated about a horizontal axis. Large, heavy workpieces will deflect the spindle, causing a loss in accuracy and precision.

30.   On a horizontal boring machine, the workpiece does not rotate. In other words, workpiece rotation limits the number of surfaces which can be machined in a single setup. Thus, horizontal boring machines are more flexible. This machine was one of the first to be converted to NC. On this machine: (1) Several types of machining operations can be performed with a single setup, and (2) the workpiece does not move, thus it is easy to clamp and hold large workpieces. Finally, chip disposal is easier than on vertical spindle machines (for boring blind holes for example).

31.   Figure 23-5 shows an example of work held in a fixture mounted on a face plate.

32.   The workpiece in Figure 23-6 is being held in a three-jaw chuck.

33.   Figures 23-1, 8, 10, 26, 32, 33, and 39 show a dead center.

34.   Figures 23-8 and 34 show a live center.

35.   a) A 3-jaw chuck is shown in Figures 23-1, 4, 6, 10, 13, 14, and 36.
       b) Collets are shown in Figures 23-26, 17, 31, 37, 38
       c) Faceplates are shown in Figures 23-5, 32, 39
       d) 4-jaw chucks are shown in Figures 23-1, 4, and 36

36.   Three form tools are used in this setup. The shaving tool is a form tool.

PROBLEMS FOR CHAPTER 23

1. $N = \dfrac{12\ V}{\pi\ D} = \dfrac{12 \times 200}{3.14 \times 3} = 255$ rpm

2. $CT = \dfrac{L + All}{feed \times N} = \dfrac{8 + 1}{0.02 \times 255} = 1.76$ minutes

3.   MRR $\approx$ 12 V f d = 12 x 200 x 0.020 x 0.125 = 6 cu.in./min

4.   a). Engine lathe cost TC_{EL} = (.5 x Q + .5) 18 + 0
         Turret lathe cost TC_{TL} = (.083 Q + 3) 20 + 300
         Equate $TC_{EL} = TC_{TL}$ at BEQ
         (.5 Q + .5) 18 = (.083 Q + 3) 20 + 300
         9 Q + 9 = 1.67 Q + 60 + 300
         Q = 351/7.33 = 41.9 or 42 units

      b). .5 x 18 + (.5 x 18)/41.9 = 9 + .21 = $9.21/part

5.   The feed given in the problem (for boring) is 0.5 mm/rev or about 0.02 ipr. The depth of cut is (1/2) x (89-76) or 6.5 mm or 0.255 inches. Assuming that for 1340 steel, the BHN would be in the low range (175 to 225), Figure 42-11 on page 1179 recommends a cutting speed of 80 sfpm or 24.4 m/min.

Drilling RPM for 18 mm drill = (24.4 x 1000)/(18 x 3.14) = 431
Drilling RPM for 76 mm drill = (24.4 x 1000)/(76 x 3.14) = 102.2
Boring RPM = (24.4 x 1000)/(89 x 3.14) = 87.3
Drilling time for 18 mm drill = (200 + 18/2)/(431 x 0.25) = 1.94
    min.
Drilling time for 76 mm drill = (200 + 76/2)/(102.2 x 0.64) =
    3.64 min.
Boring time = 200/(87.3 x 0.5) = 4.58 min.
Center drill time = 0.5 min.
Four changes of speed and tool settings require 4 x 1 min = 4
    min.

Total time, neglecting setup = 14.66 min. The operator actually
ran the job at 110 sfpm (34 m/min) and so he did the cycle in
11.7 min. or about 20% faster than you had estimated. It is very
easy for operators to alter the cutting parameters to either
increase or decrease the time to do a job by very large percent-
ages. Industrial engineers should be very wary of setting rates
under such circumstances.

6.   a). Replotting the data on log-log paper helps improve the
accuracy of the estimates from the NC curve.
        At Q = 10, TC/Q $\frac{1}{4}$ 20
        At Q = 100, TC/Q $\frac{1}{4}$ 8

        FC/10 + VC = 20   or   FC + 10 VC = 200
        FC/100 + VC = 8   FC + 100 VC = 800

        Therefore, 90 VC = 600 , VC = 6.66
              FC = 200 - 6.66 x 10 $\doteq$ 133.33
        b).   The NC lathe is the most economical between quantities
of about 20 to about 800. These are the Break Even Quantities
(BEQ's). Below 20 units, one would use the engine lathe. Above
800 units, use the single spindle automatic.
        c).   The NC lathe cost per unit approaches its variable
cost for very large quantities. This is because the total cost
equation: TC = Fixed cost + (Variable cost x Quantity) is being
divided by quantity, so that:
        TC/Q = FC/Q + VC or TC/Q   VC for FC --> 0
        d).   When this type of data is plotted on Cartesian coor-
dinates, it becomes very difficult to identify the location of
the BEQ's (see figure below). However, this is the real nature
of cost per unit versus quantity for processes. These curves
display a very sharp change from almost vertical to almost horizontal. Once you have passed the place where the curve goes
horizontal, you will be making very modest increases in cost per
unit no matter how big the quantity becomes. These curves show
us the true nature of unit cost of manufacturing processes.

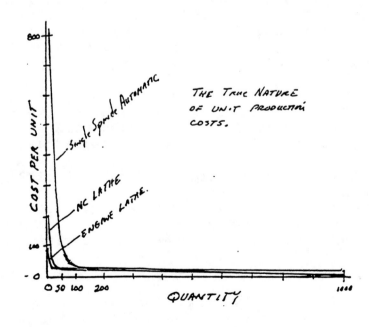

e). The elimination of setup time and thus setup cost from the above equation for unit cost means that many manufacturers have found a way to make cost per unit essentially equal to the variable cost. Graphically, for the curves plotted, it means that the steep vertical portion of the curve is eliminated. In other words, it becomes economical to build in very small lots as it does in lots of 1000 for any of their production processes. They have designed their production system to build the smallest possible lot sizes to eliminate inventory.

7. The derivation of the approximate equation 23-5 for the MRR for turning requires an assumption regarding the diameters of the parts being turned. The derivation is:

$$MRR = 12 \ (\ D_1^2 \ - D_2^2 \ ) \ f_r \ V \ / \ 4 \ D_1$$
$$= 12 \ \left(\frac{D_1 \ - \ D_2}{2} \right) \left(\frac{D_1 \ + \ D_2}{2 \ D_1} \right) \ f_r \ V \approx 12 \ V \ f \ t$$

where $(D_1 - D_2)/2 = t$ and
$$(D_1 + D_2)/ \ 2 \ D_1 \ = (D_1 + D_1 - 2t)/ \ 2 \ D_1 \ = 1 \ -t/D_1$$
$$\approx 1 \ \text{ for } \ t/D_1 \approx 0$$

which assumes t, the depth of cut, is small and negligible compared to the uncut diameter, D_1, so that $t/D_1 \approx 0$.

CASE STUDY - CHAPTER 23
Estimating the Machining Time for Turning

The cutting time for a pass over the log is:

CT = (L + Allowance) /(feed x N)
    where Allowance is in inches, feed is in inches per revolution, and N = 12 v /(π D) and is in revolutions per minute

Total time = CT x No. of passes + tool change time
    where No. of passes = ((10 - 6)/2) / depth of cut.

Thus, the real problem is to determine the maximum depth of cut. This depth of cut can be restricted (constrained) by many things, but most commonly by either power or deflection.

Let's see if the HP available limits the depth of cut.

Step 1
    Estimating speed, V, from $VT^n = C$
        For $VT^n = C$, $85 \times 10^n = 60 \times 100^n$
            $85/60 = 10^n$ $n = 0.15$, $C = 120$

    For $V \times 30^{0.15}$ $= 120$, $V = 120 / 1.665 = 72$ sfpm

    This will yield about 30 minutes of tool life (an upper limit on speed).

<u>Step 2</u> Estimate r_c for finding ϕ

20/80 as 0.4/0.6; therefore, r_c = ((0.2/60) x 52) + 0.4
= 0.57

tan ϕ = 0.57 cos 10 / (1 - 0.57 sin 10) = 0.55/ 0.90 = 0.61
ϕ = 31 or 32 degrees

HP_m = (HP_s x MRR x CF) / Eff for turning
where HP_s is estimated based on τ_s = 125,000 psi

HP_s = 2 from Table 21-3.

50 = (2 x 12 x 72 x 0.02 x depth of cut x 1.25)/ 0.75

depth of cut = d/c = (50 x 75)/(24 x 72 x 0.02 x 1.25)
= 0.86 inches

This d/c is very large, so <u>power is not the limiting factor</u>.

Let's see if deflection limits the depth of cut. The appropriate equation is:
d/c = (F_s sinϕ) /(f_r x τ_s) and requires an estimate of F_s

So F_c and F_t are needed. F_R causes deflection.
For a beam 6 inches in diameter, supported between centers, deflection at mid-beam, δ , for a round beam =>
δ = (W 1^3)/(48 E I).
where F_R = W and I = πD^4 /64 = 0.049 D^4

Therefore:
$$0.005 \text{ inches} = \frac{F\ (8 \times 12)}{48 \times (30 \times 10\ ) \times (0.49 \times 6\ )}$$
So, F_R = 522 pounds.

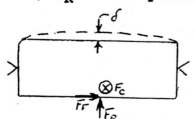

So, F_f = 1044 and F_c = 2088

F_s = F_c cosϕ - F_t sinϕ
= 2088 cos(32) - 1044 sin(32)
= 1210 pounds

Since τ_s = F_s /A_s = F_s sinϕ/ (feed x depth of cut),
Depth of cut, d/c = 1210 x sin(32) / (0.02 x 125,000)
= 0.25 inch

<u>Deflection limits the depth of cut to about 0.025 inches.</u>

The number of passes, now becomes:
N = ((10 - 6)/2) / 0.25 = 8

Cutting time, $CT = (L + All) / (\_f \times (12\ v/\pi D))$
$= ((8 \times 12) + 0.5) \times 3.14 \times \overline{D}/ (0.02 \times 12 \times 72)$
for one pass

$\overline{D}$ = average beam diameter = 8 inches

Therefore, the total cutting time would be:
$((96.5 \times 3.14 \times 8) / (0.02 \times 12 \times 72)) \times 8$ passes

For nine passes (8 at 0.25 in. and one finish cut), the cutting time would be 1262 minutes.

Tool change time would be $(1262/30) \times 2 = 84$ minutes

Total machining time would be = 1262 = 84 = 1346 minutes = 22.4 hours

For each roller, it will take about one day (three shifts) with about 42 tool changes (about 14 per shift).

COMMENT: This case study is typical of the kinds of estimates and calculations needed to estimate cutting time for heavy machining operations where only partial information is provided.

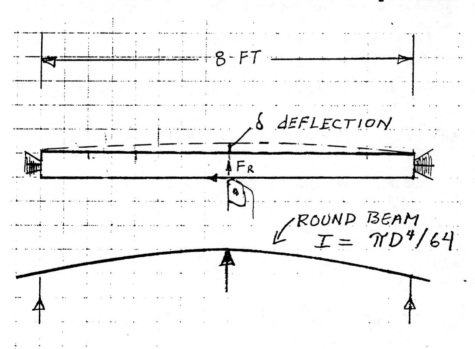

SUPPORTED AT BOTH ENDS
LOAD AT CENTER

$$\delta = W\ell^3 / 48EI$$

$$E = 30 \times 10^6 \text{ for STEEL}$$

Chapter 24

DRILLING AND RELATED HOLE-MAKING PROCESSES

1.   The flutes form the rake angle of the cutting edges, permit coolant to get to the cutting edges, and serve as channels (elevators) through which the chips are lifted out of the hole.

2.   The rake angle of the drill is determined by the helix angle of the drill at the outer extremities - the tips - and gradually changes to a zero rake angle at the inner extremities -the chisel edge. The center core drill shown in Figure 24-4 on page 702 has a small, uniform rake angle.

3.   The helix angle is mostly determined by the material being drilled.

4.   The smaller hole provides a guide for the cone portion of the point of the larger drill, of sufficient size so the chisel point of the latter does not contract the workpiece at the start and thus cannot cause the drill to wander. In addition, larger drills can drill faster if the central region of the hole is drilled out first as the negative point is removed from the operation, lowering the cutting torque and thrust considerably. Of course, an extra operation is needed (i.e. drilling the smaller hole first).

5.   <u>Area of the hole</u> times <u>the feed rate</u>, where f_rN is the feed rate.

6.   Spade drills typically are operated at slower speeds (lower rpms) and higher feeds than twist drills.

7.   The hole will generally be oversize as the drill will not be cutting properly and will probably use more torque and thrust.

8.   The drill selected to machine the hole generally has a diameter equal to the nominal hole size, so unless the drill has excessively worn, the hole will typically be equal to the nominal size or greater.

9.   Two primary functions of a combination center drill are: (a) To start the hole accurately at the desired location, and (b) to provide a tapered guide for the drill to be used.

10.   The margins bear or rub against the drilled hole and help to guide the drill and prevent it from bending. This rubbing action also produces heat which expands the drill and increases the rubbing and friction which can increase the torque. Proper lubrication is advised to reduce the friction at the walls of the hole.

11.   Drift is a particular problem with small drills and deep holes. If the rake angles or the lengths of the cutting edges between the two sides are not equal (this is usually due to improper regrinding of the drill), a force imbalance can cause the drill to drift off line. Hard spots in the workpiece can also cause the drill to move off line, as can a large void or other material nonhomogeneities.

12.   Such drills are usually employed for long, deep holes. Check Figures 24-12 and 24-21.

13.   The deeper the hole, the greater the surface area of the drill in contact with the hole wall (the margin) and in contact with the chips coming out the flute. The chips can also pack in the flute and increase the friction and thus the torque.

14.   Cutting fluids have lubricants to reduce the rubbing friction between the drill margins, and the chips, as they contact the walls of the hole.

15.   There are two drills in Figure 24-12 with oil holes, and one in Figure 24-21. Notice all these drills have straight flutes.

16.   A gang-drilling machine has several independent spindles mounted on a common base, and usually has a common table. A multiple-spindle drilling machine has several spindles driven and fed in unison by a single powered head.

17.   The thrust force (the force 90° to the cutting force or torque) increases with increasing feed. See Figure 24-4.

18.   Holding the workpiece by hand may result in broken hands, fingers, or even arms as the workpiece may catch on the drill, particularly at breakthrough, causing the workpiece to rotate at the drill rpm.

19.   Centering insures that the drill will start at the right location and not walk off the desired spot. Drilling creates the hole itself. Boring produces a sized and properly aligned hole over the entire length, correcting for any drift problems. Reaming provides for final finish and exact hole size.

20.   The slot-point drill reduces the thrust significantly compared to other drills by eliminating the chisel end of the drill. The material in the center of the hole is left undrilled and is periodically fractured away as the drill advances. See Figure 24-4.

21.   Spot facing produces a smooth surface normal to the hole axis, as a bearing surface, usually for a bolt head, washer, or nut.

22.   Counterboring produces a second hole of larger diameter and with a smooth bearing surface as its bottom, which is normal to the axis of the hole. See Figure 24-22.

23.   Reaming provides for excellent hole finish and more exact size.

24.   Shell reamers are cheaper, because the arbor is made of ordinary steel and may be used with more than one shell. Only the shells are made from HSS or coated HSS.

25.   First, the geometry of a drill for plastics will be very different than a drill for cast iron. It will have much larger helix angles and therefore larger rakes. Also, plastic is a very poor heat conductor compared to cast iron so the frictional heat will remain in the drill, causing it to overheat.

26.   The drill should be withdrawn from the hole at frequent intervals to remove the chips and permit the drill to cool. This procedure is called pecking. Ample coolant should be used. See also Figure 29-15.

27.   A spade drill requires a much smaller amount of the expensive cutting tool material, and it can be made more rigid than a comparable twist drill. Also, the different point geometries allow these drills to start more accurately. They are really more like milling cutters than drills, and are used for large holes that are not too deep (there are no flutes to carry the chips out of a deep hole).

28.   The drill bit is repeatedly withdrawn from the hole during the drilling process in order to clear the flute of chips. This procedure is envoked whenever the hole depth to drill diameter exceeds 3 to 1. See also Figure 29-15.

29.   Recall the equation which relates rpm to cutting speed:
$$V = \pi D N / 12$$
Write the N term as the ratio of drill rate (in./min.) divided by the feed rate (in./rev.)
$$N = f_m / f_t .$$ Therefore, $V = \pi D f_m / (12 f_t)$
In order to keep the cutting velocity at the drill tips constant (keep V constant), while maintaining the same penetration rate (keep f_m constant), the feed rate must increase in proportion to the drill diameter, D.
$$D/f_t = 12 V/ (\pi f_m)$$

30.   If the feed is too large, one could experience drill fracture up the middle of the drill, chipping of the cutting edge, and rough walls on the drilled hole. See Table 24-7.

PROBLEMS FOR CHAPTER 24

1.   The selection of proper speeds and feeds is the first step in any process analysis or planning. Someone has to decide the cutting parameters. Since this is an indexable-insert drill, Table 24-1 can be used. Otherwise, standard references like the Machinability Data Handbook can be used. For 1020 cold rolled steel, the recommended speeds and feeds are 400-550 sfpm and 0.004-0.007 ipr respectively. The allowance would typically be D/2. You might want to use a spade drill here as it is less expensive and ideal for shallow, large-diameter holes - See Figure 24-13.

2.   Problem 1 is solved here using a cutting speed of 410 sfpm and a feed of 0.005 ipr. The allowance used is one half of the drill diameter.

$$CT = \frac{\text{Hole depth} + \text{Allowance}}{f_r \, N} = \frac{2 \;+\; 0.75}{0.005 \times \left(\frac{12 \times 410}{3.14 \times 1.5}\right)}$$

$$= \frac{2.75}{0.005 \times 1044.6} = 0.526 \text{ min.}$$

3.   Cutting speed = 200 fpm
$N = (12 \times 200)/\ 3.14 \times 1.5\ = 509$ rpm
$MRR = \pi\, D^2\ /4 \times N\, f_r$
    $= (3.14 \times 1.5^2)/4 \times 509 \times 0.010$
    $= 1.76 \times 509 \times 0.010$
    $= 8.99$ cu.in./min. or 9 in^3/min

4.   HP = 0.9 x 9.0 = 8.1 horsepower

5.   CS = 200 fpm   1hp = 0.7457 kW
     N = 509 rpm   so 2hp = 1.5 kW
    $MRR = \pi\, D^2\ /4 \times N\, f_r$ See Table 10-2, p.248
       $= (3.14 \times 2^2)/4 \times 509\, f_r$
    Also, $HP = HP_s \times MRR$
    So, $MRR = HP/HP_s\ = (2 \times .75)/.70 = 2.14$
    Therefore, $2.14 = 3.14 \times 509\, f_r$
       $f_r = 0.0013$
       f_r (max) = 0.0013 ipr
    The process is severely limited to light feeds.

6. $MRR = (\pi\, D^2\, L/4)\ /\ (L\ /\ f_r N)\ =\ (\pi\, D^2\ /4)\ (f_r N)$
     $=\ (\pi\, D^2\ /4)\, f_r\ (12\ V/\pi\, D)$
     $= 3\, D\, f_r\, V$

7.   Yes. $f_r N$ = feed rate in inches/minute

8.   The time to change the drill is spread over the total number of holes drilled between tool changes. The units are time per hole.

9.   Spade drill:
            Feed rate = 204 x 0.009 = 1.836 in./min.
            Holes/min = 1.836 / 3 in/hole = .612
            Cost/hole = (45/60)/.612 + 160.90/ X
                    (where X = holes/tool)

        Indexable-insert drill
            Feed rate = 891 x 0.007 = 6.237 in./min.
            Holes/min = 6.237 / 3 in/hole = 2.079
            Cost/hole = (45/60)/2.079 + 285.80/ X

        Equating:
            1.225 + 160.90/X = 0.36 + 285.90/X

        BEQ:
            X = (2.85.80 - 160.90)/(1.225-0.36)
              = 124.90 / 0.865
              = 144

    If you were doing more than 144 holes, the extra cost of
the indexable-insert drill may be justified.

    Why is it a "reasonable assumption" to assume that both
tools make the same number of holes? The cutting speeds selected
here are from tables of recommended speeds, and these tables
typically recommend speeds that give 60 minutes of tool life.

    Of course, the decision to change from one process to
another is made on the basis of many factors in addition to tool
cost.

10. a). The tolerances between the holes is based on ± 1 degree
        Converting degrees into inches, ± 1 degree = (3.14 x 6)/360
        =   0.05 inches

    b). Yes, a multiple spindle drill setup can meet this
    tolerance specification, as such a setup would have a
    process capability for hole location of þ 0.030 to 0.050
    inches, assuming good drills are being used.

    c). Using a drill jig would improve the situation by an
    order of magnitude to ± 0.003 to ± 0.005 inches.

1).   The machining difficulty was in starting the drill for
machining the bolt hole on the inclined, rough cast iron surface.
The drill point would tend to "walk" down the surface, and drill
breakage would result due to bending of the drill bit.
2).   The failures were in the form of cracks, as shown at (a)
below. These failures were caused by sharp corners (produced by
the counterboring operation) which created a stress concentration
(at X) when the leg was placed in service and received a moment
bending load. The sharp corner was placed in tension, and cast
iron is weak in tension. NOTE: This problem could be further
aggravated if white cast iron were produced in this region as a
result of rapid cooling rates in the thinner leg segments.

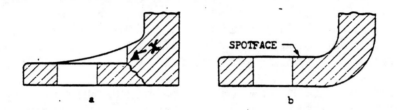

a   b

3).   One solution would be to redesign the part as shown in (b)
to eliminate the counterbore and sloped surface from the processing. Another alternative would be to use a start drill, drill
and counterbore sequence with the counterbore enlarged somewhat
and given a large radius to eliminate the sharp corner at X. If
possible, drilling from the bottom would eliminate the drill
"walking".
4).   To stop failures in the field, it will be necessary to
eliminate the sharp corner. The same oversize counterbore
operation can be manually performed on units in the field if they
have not been installed. Those that have been installed and
failed must be replaced since cast iron cannot readily be
repaired. The company should replace these at no cost to the
customer. Other installed units should be repaired (and/or
replaced) as rapidly as possible, particularly if failures in the
field could lead to personal injury.

Chapter 25

MILLING

1.   Overall, milling tends to be a faster process (shorter cutting time, greater MRR) than shaping because it uses multiple tooth cutters, higher cutting speeds, and often can complete the surface in one pass.

2.   In peripheral milling, the surface is generated by teeth located on the periphery of the cutter (See Figure 25-8). In face milling, the generated surface is at right angles to the cutter axis (see Figure 25-2).

3.   If the volume of metal removed is the same and the only difference is the direction of rotation, one would think the power (FcV) would be the same. In climb milling, a component of the cutting force is in the same direction as the feed force which lowers the power requirements on the feed motor.

4.   Casting surfaces can be quite hard (due to rapid cooling) or contain hard spots (rapid cooling around grains of sand) as well as abrasive grains (in sand castings). These factors can lead to more rapid tool wear when the cutter tooth comes down into the surface from above versus from below as in up milling.

5.   Milling cutters can be classified as arbor mounted or shank mounted. Another method is by tooth relief -- profile and form or cam relieved.

6.   You would have end milling, and you would be milling a slot.

7.   Helical toothed cutters enter the workpiece progressively. Thus, the impact of initial tooth contact is less, and, overall, the cutting forces are smoothed out as two or three teeth are engaged in the work at the same time.

8.   Imagine all the Fc patterns superimposed on each other which forms a steady Fc with small scallops.

9.   To produce a T-slot by milling: (1) Machine (using an end mill) a vertical groove sufficiently wide to clear the shank of the T-slot cutter to the full depth of the T-slot, and (2) complete the wide portion of the slot with the T-slot cutter.

10.   The teeth are staggered so that the teeth can be given a side rake angle in addition to a back rake. This reduces the impact at entry and the cutting forces overall.

11.   The table on a plain column-and-knee milling machine cannot
be swiveled to permit cutting a helix, as required for the flutes
of a twist drill. A special attachment to hold a ball end mill
and a universal dividing head can be used as shown in Figure
25-17.

12.   The turret type milling machine is more flexible than an
ordinary vertical spindle milling machine because its tool-
holding head can be swiveled about the vertical axis.

13.   In a universal milling machine, the table can be swiveled
about the vertical axis.

14.   The table can move horizontally, left, and right. The
table sits in a saddle, which can move horizontally, in and out.
The saddle sits on the knee which can raise and lower the
saddle-table assembly.

15.   The rate ring limits the amount of deflection of the
stylus. See Figures 25-14 and 25-15.

16.   Bed type milling machines have a more rigid construction,
permitting heavier depths of cut to be made.

17.   A duplicator can reproduce three dimensional shapes while a
profiler can only work in two dimensions. Both types of machines
have largely been replaced by NC or CNC machines, which can do
both duplicator and profile work.

18.   Planer-type milling machines have replaced planers because
they have one or more multiple-edged cutters and a complete
surface can be machined with one reciprocation of the table.

19.   The vertical spindle Bridgeport was (and still is) a very
versatile machine, and it was very accurate and precise.

20.   The dividing head uses a worm-gear reduction assembly. See
Figure 25-17. When you turn the crank one revolution, the
spindle on the other end rotates 1/40 of a revolution. The index
plate is designed such that a workpiece can be rotated through
almost any desired number of equal arcs.

21.   Connecting the input end of a universal dividing head to
the feed screw of the milling machine causes the workpiece to be
rotated a controlled amount as the table moves longitudinally.
See Figure 25-17.

22.   The hole-circle plate (or the index plate) is to control
the rotation of the workpiece through a desired angle. See
Question 23.

23.   The only hole circle that can be used is the 27-hole
circle. All others do not give a whole number of holes. The
calculation is (40 x 27)/18 = 60 holes. Thus a tooth gap would

be milled and then the gear blank rotated by cranking through 60 holes or about 2.22 revolutions of the crank, and then the next tooth gap milled. The setup is shown in Figure 31-7 for cutting helical gears.

24.   As in all machining processes, the cutting speed (V) is selected based on the cutting tool material and work material. The feed (f_t) is also selected in terms of how much each tooth will remove during each pass over the work -- the feed per tooth (See Table 25-1). The RPM is computed from the selected speed by:
        $N = (12 \times V)/(3.14 \times D)$ where D is the cutter diameter
Then, the table speed is calculated from:
        $f_m = f_t \times n \times N$ where n is the number of teeth in the cutter.

25.   See answer to question 24

26.   Down milling is being performed because of the geometry of the part with respect to the cutter. Besides, the part material here is likely to be cast iron, a brittle material. The top surface has already been machined. Therefore, there will be no problem down milling the large groove.

27.   The large cutting forces in slab milling must be considered. These forces tend to dislodge the part in slab up milling. In vertical spindle milling, the chip engagement (chip thickness) tends to stay more uniform and overall the cutting forces are not as large.

28.   A cutting speed of 50 to 100 fpm and a feed per tooth of 0.005 to 0.010 ipt are quite reasonable values. The student must go to a handbook or similar source to find the values. See also Table 25-1.

PROBLEMS FOR CHAPTER 25

1. $V = \pi D N/12$, so $N = 12 V / (\pi D) = 12 \times 200 /(3.14 \times 8)$
      $= 95.5$ rpm

      $f_m = n N f_t = 10 \times 95.5 \times 0.01 = 9.55$ ipm

2. $N = 12 \times 70 / (3.14 \times 6) = 44.5$ rpm

      $f_m = n N f_t = 8 \times 44.5 \times 0.012$ ipt $= 4.28$ ipm

      $CT = (L + All)/f_m = (12 + 3 + 3) / 4.28 = 4.2$ min.

3. $MRR = Vol / CT = W t f_t = 5 \times .35 \times 4.28 = 7.49$ cu.in./min.

4. $HP = MRR \times HP_s = 7.49 \times 0.67 = 5$ horsepower

5.   No. In milling the two sides could easily be machined simultaneously by stradle milling. These probably would be done first, followed by milling the bottom, then the end. Shaping would be done one surface at a time, probably starting with the large bottom surface.

6.   The first step in the problem is the selection of a cutting speed. From Table 25-1, the student might select anything from 40-130 sfpm. Let's say 120 sfpm is selected.

For face milling:
    RPM of cutter = (120 x 12)/(3.14 x 8) = 57.3 rev/min
    Table feed, f_m = n N f_t = 10 x 57.3 x 0.010 = 5.73 in/min
    CT = Machining time where A = D/2
       = (L + A)/f_m = (18 + 4) /5.73 = 3.83 min/part
    Setup time (a one time operation)   =   60.0 min.
    Load and unload fixture (very conservative) = 2 min
    Total time for one part is   =   65.83 min

    Cost to make one = $\underline{60.00}$ x $\underline{33.25}$ + $\underline{33.25 \text{ x } 5.83}$ = $37.03/part
                        60   1   60

    Cost to make 10 = $\underline{60}$ x $\underline{33.25}$ + $\underline{33.25 \text{ x } 5.83}$ = $6.55/part
                       60   10   60

    Cost to make 100 = ...   = $3.56/part

For shaping, use V= 120 sfpm (high but used for comparison):
    V = 2 l N_s / 12 R_s, where R_s = 5/9, l = 6 (Equation 26-8)
    N_s = 66.6 This is RPM of the bull wheel (See Figure 26-32)
    CT = 18/ (66.6 x 0.015) = 18 min/part (Equation 26-10 with no
                                                         allow.)

    Cost to make one = $\underline{10}$ x $\underline{25.25}$ + $\underline{25.25 \text{ x } 18.00}$ = $12.77/part
                        60   1   60

    Cost to make 10 = ...   = $7.99/part

    Cost to make 100 = ... = $7.61/part

    The shaper is cheaper when the lot size is very small. At some higher number of parts, the milling machine will be the better choice. Note that the reduction or elimination of setup time could make milling the choice even for a lot size of one.

7.   For milling, the percentage of time spent in nonmachining activities is (60 + (2 x 10))/((5.83 x 10) + 60) = 80/118.3 = 67.6%. For shaping, the percentage of time spent in nonmachining activities is (10 + (2.0 x 10))/((18.0 x 10) + 10) = 30/191 or 15.8 percent.

8.   N = 12 V / π D = 12 x 125 /(3.14 x 5) = 95.54 rpm

      f_m = n N f_t = 12 x 95.54 x 0.006 = 6.88 ipm

9. $MRR = W\ d\ f_m = 2 \times 0.5 \times 6.88 = 6.88$ cu.in./min
    Up milling is shown in Figure 21-5.

10. $N = 12\ V\ /\ \pi\ D = 12 \times 500\ /(3.14 \times 6) = 318$ rpm

    $f_m = n\ N\ f_t = 8 \times 318 \times 0.010 = 25.46$ inch/min.

    $CT = (L + \text{Allowances})/\ f_m$ where $\text{Allow} = \sqrt{.35(6 - .35)}$
      $= (12 + 1.4)\ /\ 25.46$
      $= 13.40\ /\ 25.46 = 0.52$ min

This cutting time is considerably less due to the high cutting speed for carbide cutting tools. The MRR is greater than for face milling with HSS tools.

Chapter 26

BROACHING, SAWING, FILING, SHAPING AND PLANING

1.   The feed is built into the teeth of the broach -- the rise per tooth is the feed. It is also as close to orthogonal machining as one finds in industry.

2.   The saw blade has no "step" or rise per tooth between successive teeth, so a saw blade is not a broach.

3.   These machines use straight line movement and the feed is built into the tool, so the machine tools are much simpler, mechanically speaking.

4.   Why broaching is suited for mass production -- Accuracy and precision are built into the process. No machine adjustment is needed after the initial setup. The rapid, single stroke or one pass completion of parts leads to easy A(2) or A(3) levels of automation. Roughing and finishing are built into the same tool.

5.   The pitch or the distance between each tooth. This is needed to determine how long the broach must be to remove the material. See question 6.

6.   Because all metal removal (depth of cut) is built into the tool, the design of the tool must relate to the amount of material to be removed, chip thickness per tooth, tooth-spacing (pitch and gullet size), and the length of available stroke in the machine.

7.   Methods for reducing force and power requirements in broaching are rotor-tooth design, double-cut construction, and progressive-tooth design.

8.   The rotor-tooth broach would be longer.

9.   In designing a broach, the distance between the teeth (the pitch) and the shape of the gullet (the radius) must be such that the chip can be fully contained and allowed to curl properly, so that the chips do not rub the machined surface.

10.   Since the entire surface is machined in one pass, the operation is very fast without resorting to high cutting speeds. High speed would consume more power and also generate more heat, thereby greatly shortening the life of the broach. Because these tools are usually quite expensive, they must have a long life to make the cost per part low and the entire process economical.

11.   Shell-type construction reduces the cost of the broach because the main shaft can be made of inexpensive steel, and also the shaft can be used with various sizes and types of shells. Also, worn or broken teeth can be removed and replaced and the entire broach does not have to be replaced.

12.   Because the cutting speeds are low, carbides are not
needed. In addition, the cutting forces tend to put the broach
tooth geometries in tension, where carbide is not as strong and
reliable as steel. Carbides and ceramics can be used for the
burnishing rings (i.e. finishing teeth).

13.   TiN-coated HSS broaching tools will cut with less power and
lower forces because of the lower tool/chip interface friction
condition. The lower interface friction condition produces
larger shear angles and lower shear forces. The TiN-coated tools
also last longer.

14.   It is easier to feed pull-up machines, and the work falls
free after the operation is completed.

15.   There would be no way of getting the set of broaches into
the hole.

16.   No. There would be no place for the chips to go, and the
first tooth on the broach would have to be full size, permitting
no feed being built into the tool.

17.   Such sockets usually have a recess, larger than the
finished size of the broached hole, beyond the end of the surface
to be broached. Such recesses can be made by forging, casting,
or machining.

18.   Sawing is relatively efficient because only a small amount
of material is formed into chips.

19.   (1) Tooth spacing controls the size of the teeth, (2) the
spacing determines the space into which the chips must be
contained, (3) tooth spacing determines how many teeth are in
contact with the work (cutting) at a given time. Tooth spacing
is the same as pitch in broaching.

20.   The tooth gullet is the space between the teeth. It must
be large enough to hold all the chips from a single pass over the
workpiece.

21.   "Set" is the manner in which the teeth are offset from the
centerline of the saw blade so as to produce a cut that is
slightly wider than the thickness of the blade. The width of the
cut is called the "kerf". See Figure 26-16 and 26-18.

22.   If the band were hardened throughout its width, it would be
brittle and would break when flexing around the guide wheels.

23.   The wider the blade, the larger the minimum radius of a cut
that can be made. See Figure 26-19

24.   Circular saws are limited in the depth of cut that can be made with them. Also they are more expensive than bandsaws. Advantages: they can be made stronger, more accurate cuts can be made, and they have teeth made from a variety of cutting materials.

25.   Bandsawing machines can operate at higher cutting speeds and cut continuously (no reciprocating) and are thus able to make the same cuts faster than hack saws.

26.   A hole is drilled into the workpiece. The bandsaw blade is broken, inserted into the workpiece, and welded. The cuts (holes) are then made. The blade is broken and removed. This process is good for small volumes of parts.

27.   In friction sawing, there are no teeth to form chips in the normal fashion. The friction from the continuous rubbing of the tool on the work heats the metal in the cut and softens it. The blade then rubs or pushes it out of the kerf.

28.   If feed is by gravity, the feed force is constant. As the cut proceeds, the length of the cut increases. The force resisting the feed increases in proportion to the length of the cut. Thus, the feed rate slows down and speeds up in proportion to the diameter of the round bar.

29.   The file is much wider than the saw blade and the teeth may have negative rakes, but these are the only real differences.

30.   A safe edge on a file means that the file has no teeth on the edge. The user is less likely to be injured while using it and metal won't be filed from undesired locations.

31.   On a band filing machine, the cutting motion is continuous - i.e. no reciprocation.

32.   The teeth on a rasp-cut file are formed by being plastically deformed outward from the body of the file, whereas those of other types are formed by cutting.

33.   In a shaper, the tool reciprocates and the work feeds perpendicularly to the tool motion. In a planer, the work reciprocates and the tools feed perpendicular to the work movement. Both make straight line cuts.

        Shapers are best suited for flat surfaces on small workpieces in small quantities as in the tool room or for special one-of-a-kind jobs. Planers are used for large workpieces.

        Because the workpieces machined on planers are large and heavy, it is difficult to reciprocate the work and table rapidly and to block the workpiece so as to hold them against the high acceleration and deceleration forces occurring at the ends of the strokes.

It is good practice to cut on a shaper with as little overhang of the ram arm as possible. The arm is a moving cantilever beam and the cutting forces will greatly increase the amount of deflection in the arm as the length of overhang is increased. The planer does not have the cantilever beam design of the shaper, so it can make long straight cuts without suffering deflection problems, and therefore can take advantage of the cutting time saved.

34.   Shaper feed is in millimeters or inches per stroke, while milling is in inches per tooth. In shaping, the cutting time is relatively slow and the setups, while usually simple, can take as long as the setup on a milling machine, which will have a faster cutting time. Thus, milling is generally able to show an economic advantage over shaping and has about the same or better precision. See Problem 6 in Chapter 25.

35.   On planers, two tables are often used, so one is being used while the other is machining.

        On planers, the setups and cuts are designed so that cuts are made during both the forward and return strokes, while on shapers, cuts are made only on the forward stroke and feed occurs after the tool has returned. (On both shapers and planers, feed is in inches per stroke.) On planers and shapers, the table cannot be reciprocated at high speeds, so cutting speeds are relatively low and cutting time is large. On planers, simultaneous cuts can reduce the cutting time.

        These methods CANNOT BE USED ON A SHAPER.

PROBLEMS FOR CHAPTER 26

1.   The length of the cut times the feed is 12 x 0.0047 = 0.0564 cu.in. per gullet. This would be the minimum cross section of the gullet. The gullet would have a larger cross section than this to allow the chip to curl.

2.   The formula used to estimate the pitch is an empirical expression based on English units. Thus, the metric units must not be used. $P = .35 \sqrt{Lw} = .35 \sqrt{17.75} = 1.47$ inches. The number of teeth needed are 0.25/ 0.004 = 62.5 teeth. The length of the roughing section is then 63 x 1.47 = 92.6 inches.

3.   For gray cast iron, HP_s = 0.5 HP/ cu. in./ min.
        10 m/min = 32.75 ft/min
The horsepower needed per tooth:
        $HP = HP_s \times MRR$ = .5 x(12) (0.004) (3) (32.75)
              = 4.716 horsepower

The number of teeth in contact:
        = 17.75/ 1.47 = 12

The maximum HP is = 12 x 4.716 = 56.59 hp, a rather large value, suggesting that the broach be redesigned if the machine does not have sufficient horsepower.

4.   The approximate force per tooth can be estimated by: HP = F_c V/ 33,000. Therefore, $F_c \cong$ 4.716 x 33,000 / 32.75 = 4752 lbs per tooth. For 12 teeth, this requires 57,024 pounds. This is very large. The student should be concerned.

5.   The cutting speed selected should be around 55 m/min. = 180 ft/min. The pitch is 0.05 inch or 0.00417 ft. The number of teeth which pass over the workpiece per minute = 180/0.00417 = 43165.5 teeth/min. The CT = 6/ (43,165.5 x 0.0001) = 1.39 min. with no allowances.

6.   Allowable pull = (A_{min} x Y.S.)/S where S = factor of safety
        Allowable pull = ((πD_p^2/4 - D_pW) x 200,000)/ 1.25
              where D_pW = the area of the slot in the pull end.

7.   First determine the Stroke Ratio, R_S = 200/360 = 0.55
        N = 12 V R_S / 2 l where we let l = 2 L to allow for overrun
              at both ends of the stroke and allow the ram to reach
              full cutting speed before it enters the workpiece
        = 12 x 25 x 0.55 / 2 x 4 = 165/ 8 = 20.6 rpm or 20.6
              bull wheel strokes per minute

        Cutting time = CT = W/(N_S x f_c) = 7/(20.6 x.1) = 3.39 min.
              Note that shapers are rather slow.

        Metal removal rate = MRR = L w t / CT = (4 x 7 x .25)/3.39
              = 2.06 cu. in./min.

8. R_S = 0.55 ; N_S = 11.78 for l = 7 inches; CT = 3.39 minutes
Thus, this setup does not take less time, but requires a much greater overhang on the ram and a possible loss of accuracy and precision due to deflection.

        N_S = 12 V R_S / (2 l) = 12 x 25 x 0.55 / (2 x 7) = 11.78

        CT = w / (N_S x f_c) = 3.39 min

9.   Let $\bar{V}$ = average velocity of the ram produced by N rpm of the crank with a R_S stroke ratio.

        $\bar{V}$ = $\dfrac{distance}{time}$ = $\dfrac{length\ of\ stroke}{time\ of\ stroke}$

        $\bar{V}$ = $\dfrac{1}{(1/N_S)\ x\ R_S}$ $\dfrac{in.}{min.}$

        Since V = 2 $\bar{V}$
        V = $\dfrac{2\ l\ N_S}{12\ R_S}$ $\dfrac{ft.}{min.}$

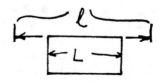

10. For $R_S = 220/360 = 0.611$
$N_S = (12 \times 120 \times 0.61)/(2 \times 10) = 43.92$ strokes/min
or $N_S = (6.11.1 \times 36.6)/(2 \times 254) = 44.03$ (metric units)
    where $611.1 = R \times 1000$ mm/meter
so, $N_S = 44$ strokes/min

11. $N_S = (12 \ V \ R_C)/(2 \ l)$ but $R_S = 2/3$ for hydraulic shapers
    with a 2:1 cut to return ratio and $= L + 1$ inch
    allowance instead of 2.
    Therefore:
$N_S = 8 \ V \ /l = (8 \times 150)/(8 + 1) = 133.3$ strokes/min
$CT = W/(N_S \times f_C) = 10/(133.3 \times 0.020) = 3.75$ min.

12. $MRR = L \ W \ t/ \ CT = (10 \times 8 \times 0.25)/(3.75) = 5.33$ cu.in./min.

13. Assuming gray cast iron has a specific horsepower of
    0.30 HP/ cu.in./min,
$HP = HP_S \times MRR = 0.30 \times 5.33$ cu.in./min. $= 1.59$ HP

14. Available power $= 7.5 \times 0.75 = 5.625$ HP
The maximum $MRR = 5.625 \times 0.67 = 3.76$ cu.in./min.
$MRR = L \ W \ t/ \ (W \ /(N_S \ f_f)) = L \ t \ N_S \ f_C = L \ t \ 8V \ f_C \ /l$ for
    hydraulics
Therefore, $t = MRR \ / \ (8 \ L \ V \ f_C)$
    $= (3.76 \times (12 + 1))/(8 \times 12 \times 180 \times 0.025)$
    $= 0.113$ inches
    where $l = L + 1$.

CASE STUDY - CHAPTER 26
The Component with the Triangular Hole

After getting over your initial reaction to "who the heck
designed this part?" and "I don't think that this part can be
made!", you would find that there are really many ways to make
the part. Clearly, it could be made by powder metallurgy or
investment casting. It could be made in two pieces -- a flat
disk and cylinder with a broached triangular hole -- with an
appropriate joining process, perhaps friction welding. The hole
could be machined into the cylinder by EDM, ECM, or ultrasonic
machining. With the latter three approaches, a hole should
initially be drilled of a diameter about 5 and 1/2 mm in the
center of the cylinder. Two EDM tools are used: a triangular
hollow tool followed by a solid triangular tool to finish the
hole to size. One might want to follow initial drilling with end
milling, to make the initial hole flat bottomed. Drilling and
milling before EDM, ECM, or ultrasonic will greatly enhance the
overall processing time.

If you can get the designer to relent a bit on the 0.8 mm
radius, you can use the Watts method of drilling angular holes.
The Watts method consists of a Watts Patented Full-floating
Chuck, Angular Drill, and Guide Plate and is kind of a Wankel
engine that machines. Triangular, square, and hexagonal holes

can be drilled on conventional lathe, mill, or drill press equipment. Again, a regular round hole is drilled first in harder metals as a lead hole, but this probably won't be necessary of aluminum is selected as the metal for the part. These tools are sold by the Watts Bros. Tool Works, Inc., Wilmerding, PA.

It may be possible to make this part by backward impact extrusion, since the material is aluminum and the part is not that large. The final selection as to which processes are most economical would likely come down to impact extrusion, powder metallurgy, and investment casting. The quantity here is quite large and all of these processes can be automated. Die life may be a problem for impact extrusion because of the small radius that will have to be placed on the punches to get those 0.8 mm corners.

Chapter 27

ABRASIVE MACHINING PROCESSES

1.   Grinding, honing, lapping, and ultrasonic machining are four processes that use abrasive grits for cutting tools.

2.   Attrition is caused by the dulling of the edges and flattening of the grits, and the glazing of the wheel surface that is caused by the abrasive wear action of the grits. The grits are pulled out of the surface of the wheel as the forces on the worn grits increase.

3.   Friability is the ability of the grits to fracture and expose new cutting edges, which results in more cutting surfaces continuously becoming available.

4.   The smaller the grit size, the better the surface finish.

5.   Both are quite hard, but aluminum oxide is tougher than silicon carbide, and is less reactive with materials. Therefore, it is the more general purpose abrasive.

6.   CBN is harder and does not react with certain work materials at the elevated temperatures of grinding (particularly steel).

7.   The common bonding agents are vitreous ceramics, plastics, rubber, and silicate of soda.

8.   Grade expresses the strength of bonding material. It controls how freely grits will pull out of the wheel -- the stronger the bond, the more difficult it is for grits to pull out of the wheel surface.

9.   Structure refers to the spacing -- how far apart are the abrasive grains. An open structure has widely-spaced grains compared to a dense structure. Either structure could use a high strength bonding material.

10.   In crush dressing, the grains in an abrasive wheel are crushed, or broken, by means of a hardened roller, to expose sharp edges and, usually, to impart a desired contour to the wheel. It is the easiest practical way to impart a desired contour to an abrasive wheel. See Figure 27-16 and 27-17.

11.   A glazed wheel is one in which the grits are worn flat and polished; whereas, a loaded wheel is one in which chip material has packed in between the grains so that the entire surface of the wheel is smooth, rather than just the tops of the grains.

12.   Grinding is a mixture of cutting, plowing, and rubbing processes, all occurring at different places at the same time. Grits with large negative rakes may just plow a groove in the surface rather than form a chip. Other grits may simply rub or burnish the surface (depth of cut very small or cutting edges very rounded or worn). The grits that are making chips do so in exactly the same manner as a single point cutting tool.

13.   In dressing a grinding wheel, dulled abrasive grains are broken (thereby exposing sharp edges) or are pulled from the wheel to expose new grains.

14.   In abrasive machining, heavy feeds and large abrasive grits are used to rapidly remove material. Cutting dominates the process but, fundamentally, it is not really different from grinding.

15.   The grinding ratio is the ratio of the volume of metal removed versus the volume of wheel lost (abrasive material used) or worn away (attrition).

16.   Feed is controlled by tilting the regulating wheel. The angle of inclination provides a force in the feed direction. The part feeds at
$$F = dN \sin \theta \quad \text{(Equation 27-1)}$$

17.   There must be spacing between the grains to make room for the chips. To a certain extent, spacing or structure, along with the grain size, also dictates the surface finish.

18.   The cutting time, CT, equals the length of cut plus allowance divided by the table feed rate, f_m, times the number of passes the wheel makes over the surface, N_p. N_p equals the width of workpiece divided by cross feed, in inches per cycle.

19.   The cutting fluid carries away the chips and keeps the workpiece and grinding wheel cool. The very high grinding speeds convert considerable energy into heat energy. The grinding area is very limited and the localized heating can easily damage the workpiece.

20.   The wheel is fed radially into the rotating workpiece.

21.   The dust resulting from grinding contains fine, hard abrasive particles which can become airborn and get embedded in the softer, moving parts of other machines, causing these parts to act as laps, which thereafter would ruin the accuracy of the machines. In manufacturing cells, where grinders are often placed near other machines, it is important to put good dust control devices on the grinders and use lots of cutting fluids.

22.   The cutting time, CT, equals the length of cut plus allowance divided by the creep feed rate of the machine, f_m. The cut is usually completed in one pass. See question 18.

23.   The purpose is to produce a surface which is free of
residual stresses or a surface in which tensile and compressive
stresses are nicely balanced. Cutting results in residual
tension, while rubbing and burnishing produce residual
compression.

24.   Wheel speed is reduced, the down feed is reduced, and
sulpherized cutting oil is used for low stress grinding. See
Figure 27-10.

25.   The larger the grains, the fewer that can be packed into a
given area, so on the average, fewer grains will contact the
workpiece during a pass.

26.   Centerless grinders are faster, have better work support,
require very little operator skill, have the possibility of
continuous infeed, give excellent size control, and can be
automated with regard to part loading and unloading. Wheel
adjustment for wheel wear can be automatic as well.

27.   The through-feed is varied by changing the tilt of the
wheel and this could be done while the process is in operation.

28.   The grinding forces are much lower than the forces used in
milling and usually are directed downward into the vacuum chuck.
Milling has larger forces, and the force may be directed up, away
from the vacuum check, as in up milling.

29.   The typical grinding operation makes many passes at very
small depths of cut and relatively large feeds. In creep feed
grinding the depth of cut is large, the feed is very slow, and
the cut is often made in one pass over the workpiece. Creep feed
grinding is grinding at very slow feed rates. See Figure 27-17.

30.   In lapping, the abrasive grits become embedded in the soft
material of the lap. This is referred to as "charging the lap".
The material to be lapped is machined or rubbed by the abrasive
grits, not the soft material of the lap.

31.   In honing stones, additional materials, such as sulfur,
resins, or wax are added to the bonding agents to modify the
cutting operations. The grits themselves are very fine or small.

32.   "Charging" a lap is loading it up with abrasive materials.

33.   Honing is intended to smooth and size the hole, not to
alter the position or angle of the axis, so a rigid setup is not
what is desired in this tool.

34.   In most coated abrasive belts, the abrasive grains do not
pull out so as to expose new, sharp grains. They thus have no
self-sharpening action.

35.   Wheel wear is related to the MRR by the G ratio, which may not be linear.

36.   The angle is a function of the rate of rotation (rpm) of the honing head versus the rate of oscillation.

37.   Four major causes of grinding accidents are:
    a). operating at too high of a rpm.
    b). operating a wheel that has been dropped or struck
        so as to produce a crack
    c). operating the wheel improperly
    d). operating the wheel with the safety guards removed.

38.   A surface grinder resembles a horizontal spindle milling machine.

PROBLEMS FOR CHAPTER 27

1.   a) The wear of the stairs is produced by the hard particles of material embedded in the soles of people's shoes. The leather or rubber soles are softer but act as laps, charged with fine grits of abrasives. The bottom of the stairs are nearest the outside. b) The grits get dull as the stairs are ascended so less stair wear occurs at the top than the bottom. Soles get charged while walking outside the building, not inside the building.

2.   The small particle will tend to have fewer defects (dislocations) per unit volume and will thus act stronger. The small particle may also be more work hardened than the bulk material.

3.   In surface grinding, the MRR is controlled by the table feed, V_w. See Table 27-5.

    For a 1-inch wide wheel removing 0.004 inches of metal, the MRR = (12)(150)(0.004)(1) = 7.2 in^3/min. if the entire face of the wheel were engaged. However, the wheel is crossfed over the workpiece at a rate of 0.060 in per pass, so the
MRR = 7.2 x 0.060 = 0.43 in^3/min.

    Generally speaking, MRR's in grinding are an order of magnitude less than other multiple-tooth machining processes.

Several possibilities exist here:
One method might be to purchase tubing with the correct internal
diameter or wall thickness (if possible) and slice the rings off
the tube with a sawing or cutoff operation. This would be
followed by a milling operation to cut the opening in the ring.
A tumbling operation might be needed to eliminate sharp edges and
burrs on the ring. If tubing of the proper size cannot be
located and additional machining of the OD and ID are needed,
this method will not be the most economical.
A very economical procedure would be to purchase this material
(5052 aluminum half hard) in a wire form, and convert the wire
into a rectangular shape by pulling it through a device called a
Turks Head (see Figure below). A Turks Head is a roll forming
device with four rollers which form the four sides of the needed
rectangle. The wire would thus be given the 1.60 x 2.36 cross
section. Next, a round mandrel would be made. The mandrel
diameter would be something less than the 89,71 ring diameter.
The rectangular wire would be wound up on the mandrel like a big
spring, forming a continuous coil. This operation would be done
on a lathe and could provide the pulling means to pull the round
wire through the Turks Head. The mandrel, with the coil clamped
in place, is them placed in a milling machine and a slitting saw
is used to form the individual rings. Some calculations
(regarding springback) and experimentation would be necessary to
determine the correct diameter of the mandrel and the width of
the slitting saw so that, when the coil is cut, the individual
rings will come out with the appropriate diameter with the
correct opening.
If an extrusion press is available, the square wire can be formed
by extrusion, since a rectangular extrusion is fairly easy to do
in aluminum, and the dies might not be overly expensive. However,
if the die costs exceed $500, it would be advisable to go to the
Turks Head method, as Turks Heads do not cost much more than this
and can be adjusted and used for other applications at a later
time.

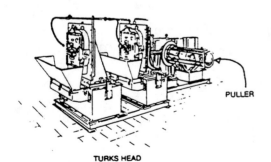

TURKS HEAD

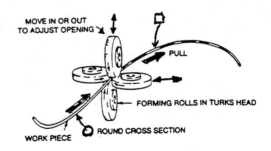

Chapter 28

WORKHOLDING DEVICES

1.   The work holding device locates the part in the machine tool with respect to the cutting tools and holds the part (clamps it) so it does not move due to cutting forces or inertial forces.

2.   A jig determines location dimensions while a fixture does not. A fixture is a special workholding device -- that is, specially designed to accomplish a specific job. Jigs have the layout of geometric shapes built into them, and thus they automatically transfer this layout to the workpiece as operations are performed with their use.

3.   The definition was incorrect, in that some jigs do not hold the work (as in clamp-on jigs), and some jigs do not guide the tool (as in welding jigs). Welding jigs are used to locate one (or more) parts with respect to another part and hold them in the right orientation and location while welding is performed.

4.   A vise is a general purpose workholding device and is not a specially designed workholding device. This answer may sound picky but it is important to distinguish between a vise used in general purpose milling and a fixture used in a milling machine. Why? The latter may have many special features designed into it to enhance or speed up production, reduce setup time, or reduce time to load or unload parts. The fixture may have a pokayoke built into it, meaning that it cannot be operated if parts are loaded into it incorrectly (pokayokes prevent defects from occurring - See Chapter 43).

5.   Some basic factors in designing jigs and fixtures are: (1) clamping the work to resist the cutting forces; (2) supporting the work during cutting so that it does not deflect under the load of the cutting forces; (3) location to provide the desired dimensional control; (4) guidance of the tool, if required; (5) provision for chip removal or clearance during or after operation; and (6) rapid, easy, safe operation.

6.   The critical surfaces (often 3 perpendicular planes) are surfaces on the part that are vital to the parts function or operation. Other surfaces are dimensioned from the critical surfaces, and these surfaces are established early in the processing sequence.

7.   The clamping forces can distort the workpiece. The workpiece is machined in the distorted configuration. When the clamping forces are removed, the workpiece returns to its unstressed shape, but now the machined surface is distorted and the dimensions produced by the machining operations will be incorrect.

8.   Three points are required to locate the workpiece in one plane. Two points are required to locate the workpiece in a second plane, perpendicular to the first plane. One point is required to locate the workpiece in a third plane, perpendicular to the first two.

9.   Supporting the work against the cutting forces of the process often requires that additional points or bearing supports be placed in the three perpendicular surfaces, beyond the 3-2-1 points.

10.   Reasons for not having the drill bushings actually touch the workpiece include: (1) Chips may become tangled in the drill bushing if there isn't sufficient clearance between the bushing and the workpiece, (2) the end of the bushing may contact an oversize workpiece and not permit the piece to be located or held properly, and (3) the chips will be passed through the bushing and may wear and score it.

11.   Down milling pushes the workpiece down into the location surfaces, which are solid, unmoving surfaces; up milling tends to lift the workpiece out of the fixture, so clamping forces must be greater to hold the workpiece against the location points during machining.

12.   The jig permits the duplication of parts without having to layout on the part the desired geometry. It is a template for locating machining operations. The use of the jig "transfers" the skill of the machine operator to the jig. The jig provides accurate and precise hole location.

13.   Forces acting against the floor caused the jig to deflect, which, in turn, caused the jig to twist. By having a rigid jig with only three points of support, the jig would not twist.

14.   If a machine is costly and has a high production rate, time lost in setting up and clamping a workpiece is very costly. Thus a small amount of time saved each cycle by use of a fixture may easily repay the cost of the fixture. A machine that is not costly or highly productive may not offer sufficient return to pay for the same fixture.

15.   Jigs that can be flipped over to permit drilling from more than one side are called roll-over jigs. They: (1) usually eliminate the cost of a second jig, (2) reduce the amount of clamping time, and (3) may reduce possible clamping error due to clamping stresses.

16.   The spherical washer permits minor deviations in the parallel surface to readily be absorbed. The strap clamp does not have to be exactly parallel to the surface holding the D stud. This allows for variations in the thickness of the workpiece.

17. Strap clamps, C clamps and toggle clamps are all commonly used.

18. The strap clamps can be bought in different sizes. The letters are used in a table in the clamp catalog to define the sizes but the basic design of the clamp does not change, just the size.

19. Rounds, buttons, and half rounds are shown in the Figure.

20. The X plane is the largest plane and would ordinarily take 3 buttons. However, this would place the thrust of the drilling process for the two mounting holes outside the area defined by the three buttons. Therefore, 4 buttons are used in the X plane and the drilling thrust is inside the region defined by these 4 points. The bottom of the bearing block would be milled flat and true prior to insertion in the jig.

21. The Z location buttons establish the "A" dimension.

22. The front and back could be straddle milled first, then the base milled perpendicular to the front or back and finally the right end. The end is milled solely for the purpose of establishing dimension "B" and "C:. The right end must rest against button"Y" in the jig. The base could be milled first and then the front and back milled, using the base as a locating surface.

23. The surfaces which locate the holes are milled first to properly establish dimensions "A", "B", and "C". It is more difficult to locate surfaces to be milled from surfaces that were drilled than it is to locate (and drill) holes with respect to milled surfaces. While the drawing does not say so specifically, the holes are perpendicular to the flat bottom.

24. Drill bushings (K) must be removable so the holes can be countersunk with the workpiece still in the jig. Drill bushings are made removable for any number of reasons. You may want to replace it if it wears. You may want to remove it so that the drilled hole can be reamed, tapped, or countersunk --the reamer, tap, or countersink being larger than the drill diameter. You may be drilling two holes of different diameter in the same location, so you need two different drill bushings.

PROBLEMS FOR CHAPTER 28

1. ($5.75 + $4.50)2.25 - ($4.50 + $4.50)1.25 =
    ($3.000/N)(1 + (3 x 0.1)/2 --> N = 292+ or 293 pieces

2.   The cost of the jig is:

$$C_t = \frac{\$100 + (4)(12.00)}{N} + \frac{\$600}{N}\left(1 + \frac{3 \times 16}{2}\right)$$

Assuming the design and assembly costs are one-time costs and the modular elements are written off over three years;

$$C_t = \$148/N + (600/N)(1.24)$$

$(8.00 + 8.75)(.5) - (6.50 + 8.75)(0.2) \cdot 148/N + (600/N)(1.24)$
$8.375 - 3.05 \cdot 148/N + 744/N$
$N = 892/\ 5.325 = 167.51$

The modular fixture has a lower breakeven quantity.

3.   The cutting force, F_C of 1800 lb. is assumed to be going to the left, and the thrust force of 900 lb. is assumed to be going down. The clamping forces, F_R and F_L are required so that the clamps can be designed.

For a static condition, the sum of forces in the X-direction (horizontal) equals zero, the sum of forces in the Y-direction (vertical) equals zero, and the sum of moments around any point is zero.

$F_x = R_1 + \mu(F_L + F_R + 1500 + 900) - 1800 = 0$

$F_y = R_2 + R_3 - F_L - F_R - 900 - 1500 = 0$

M_A = moment about point A on left side
   $= (900 \times 5) + (1500 \times 15) + (F_R \times 30) + (R_3 \times 30)$
      $- (1800 \times 34) = 0$

Let $R_3 = 0$ (assume part is barely touching) and
 $F_L = 0$ (assume there is no tendency to lift on the left side).

 So:
     if $R_3 = 0$; $F_L = 0$

     $2F_x = R_1 + 0.19 (F_R = 1500 + 900) - 1800 = 0$

     $2F_y = R_2 - F_R - 900 - 1500 = 0$

     $M_A = 4500 + 1500(15) + F_R\ 30 - 1800(34) = 0$

     $F_R = \frac{1800(34) - 4500 - 1500(15)}{30}$

        $= 1140$ lb

     $R_2 = 1140 + 900 + 1500 = 3540$ lb.

     $R_1 = 1800 - .19 (1140 + 1500 + 900) = 1127.4$ lb.

CASE STUDY - CHAPTER 28
Overhead Crane Installation

This study actually involves two location problems: the location of the holes with respect to themselves and the location of the hole patterns with respect to each other. The former problem can be solved by making a simple ring jig. The jig can be secured to the column using magnets or a hole can be drilled at the point (+) on the column, tapped, and used to hold the jig plate while the bolt holes are drilled. The holes can be drilled with a hand electric drill.

The second problem, locating the drill jig properly on each column so that the hole pattern centers all come out on the same plane, is a bit more difficult. Here is one possible solution. A fine cross (+) is placed on the jig. On the day that the job is to be done, a surveyor's transit or level is set up in the center of the eight columns at the required height. A painter's scaffold should provide adequate height. When the transit is properly leveled, each column can be "shot" so that, when the jig is mounted on each column for hole drilling, the jig will always be at the same height with respect to all other columns (without regard to the floor itself).

The equipment needed would be: a drill jig, magnet, drills, portable electric drill with long extension cord, scaffold, and surveyor's level.

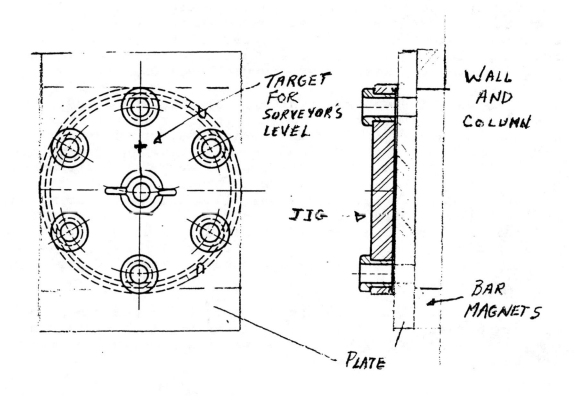

TARGET
FOR
SURVEYOR'S
LEVEL

WALL
AND
COLUMN

JIG

BAR
MAGNETS

PLATE

Chapter 29

NUMERICAL CONTROL AND MACHINING CENTERS

1.   The machining center can automatically change tools to
permit operations and processes other than just milling to take
place. Both machines can be numerically controlled. The
machining center will often be able to change pallets automatically with one pallet being in the machine and the other pallet
being outside the machine having a workpiece mounted on it. This
reduces the machine down time by doing the setup externally - the
machine does not have to be stopped during setup, only during
part exchange.

        In 1958, Kearney and Trecker marketed a NC machine tool
that could automatically change tools, thus making it a
multiprocess machine tool and the first machining center.

2.   Parsons conceived of the idea of a machine tool controlled
by inputting numbers. He demonstrated his idea to the U.S. Air
Force by having three men stand at the controls of a 3-axis
milling machine with three more people calling out numbers to
them simultaneously. The machine then produced a complex
contour. The USAF gave Parson's company a grant to develop a NC
machine. Parson subcontracted MIT and the rest is history.

3.   DNC as first practiced means direct numerical control and
described a system wherein numerical control machines where
hardwired to a large digital computer. Programs were sent
directly to the machine tool and paper tapes were not needed.
Recently, DNC stands for distributed NC were programs are
distributed to the on-board computers at the CNC machine tools.
That is, the CNCs are networked to a large computer which
provides enhanced memory and computational capacity.

4.   An adaptive control system, A(5), must be able to evaluate
the process, and modify the inputs in order to optimize the
process in some way. In order to do this, the machine tool must
have a computer, and that computer must have in it a mathematical
model which describes "how the process works". The process is
going to adapt itself to improve or optimize itself, or its cost,
or some other feature. To be specific, the house thermostat
controls the temperature in the house. To make this system A(5),
it would have to adjust fuel and air mixtures to improve the heat
yield and burn more efficiently. It may even be programmed to
change from oil to gas, depending upon the cost of fuel, in order
to optimize cost.

5.   It implies that one has a theory which describes how the
process actually works and how all the input parameters alter the
outputs. That is, it has a good working model which describes
the behavior of the process.

6.   Feedforward is sensing something about the product on the
input side of the process and altering the process to meet these
changing input parameters. For example, in hot rolling, the
temperature and size (thickness) of the plate entering the
rolling stand influences the strength and needed opening between
the rolls to accomplish the desired thickness on the output side.

7.   Perhaps the most commonly employed feedback device is the
thermostat in your house heating system or in your kitchen oven.
The toilet employs a mechanical feedback mechanism which shuts
off the water after it has reached the necessary height in the
tank. The cruise control in cars maintains the speed using
feedback.

8.   The machine tool builders had to learn how to build more
accurate and precise machine tools, removing friction and
backlash from the mechanical drives, often through the implementation of ball lead screw drivers. The machines also had to be
made more rigid, so that elastic deflections were less of a
problem. A good machinist could compensate for such problems on
a regular machine tool.

9.   Cutter offset refers to the condition where the path of
tool centerline must offset from the desired surface by half the
diameter of the tool. This means that the geometry of the path
will be different from that of the desired surface. Interpolation refers to the situation wherein paths not on the X-Y axes of
movement of the table must be approximated with a series of
connected short (X,Y) movements. The shorter the increment (the
X or Y distance moved), the better will be the interpolation but
the program will be longer.

10.   The operator performs part loading and unloading,
inspection, deburring, part transportation, reorientation (turn
part end for end), and process monitoring. Any function which
requires thinking on the part of the operator will be difficult
to automate. In addition to the above functions, the operator
may perform setups, improve setups, control the process
capability and maintain the machines. Therefore, he may be very
multifunctional. The higher the level of function, the more
difficult it will be to automate it.

11.   See response to question number 8 of this chapter.

12.   There is no reason why it cannot be open-loop. The problem
is, however, in contouring, where one must control the velocities
in the X and Y (and Z) simultaneously within a certain tolerance
or variation. This is difficult to accomplish without feedback.

13.   A shaper is not really a production-type machine and a
broach is a straight line cutting machine wherein the tool
geometry dictates the geometry of the surface to be machined.

14.   It takes too long to manually generate all the points needed to describe a contoured path, even in two dimensions. Suppose you have a one inch long curved path and the contouring requires a tolerance of 0.001 inches. This means you might have to generate 1000 sets of points to program the tool to travel one inch.

15.   The tombstone or post-type workholders used in large, horizontal spindle CNC machines are examples of modular fixtures.

16.   The feedback detection or sensor can be placed on the motor, on the ballscrew of the table or on the table itself. See Figure 29-5.

17.   The machine tool has a point in space that is its zero point - where X,Y, and Z dimensions are zero. This point is fixed in the machine tool. The zero reference point is selected by the part programmer or machine tool operator as some point on the part from which all the part dimensions are made. See Figures 29-7 and 29-8 for examples.

18.   An encoder is a feedback sensor (a device) which generates pulses as it rotates. The source of the pulses is often an interrupted light beam. See Figure 29-5 and 29-10.

19.   In continuous-path or contouring control, both velocity and position must be controlled at all times in order to keep the tool on the desired path. In three-dimensional contouring, the cutter is required to move in three directions simultaneously. That is, the movement is the resultant of X,Y, and Z components. The curved path is broken up into short straight segments or arcs.

20.   Pecking is a software routine that is already programmed into the machine which permits the drill to be periodically raised out of the hole to clear the flutes of the drill or chips. See figure below.

21.   Pocket milling is a form of contouring wherein a hole (usually square or rectangular) is milled into a solid block of metal. This is often done to reduce the weight of the finished product. The cutting tool first feeds down to the desired depth and then feeds in a contouring path to produce the pocket. End mills with spiral flutes are commonly used for pocket milling. See the following figure.

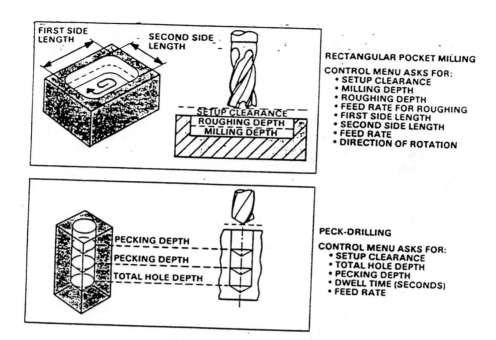

FIRST SIDE LENGTH
SECOND SIDE LENGTH
SETUP CLEARANCE
ROUGHING DEPTH
MILLING DEPTH

RECTANGULAR POCKET MILLING
CONTROL MENU ASKS FOR:
 • SETUP CLEARANCE
 • MILLING DEPTH
 • ROUGHING DEPTH
 • FEED RATE FOR ROUGHING
 • FIRST SIDE LENGTH
 • SECOND SIDE LENGTH
 • FEED RATE
 • DIRECTION OF ROTATION

PECKING DEPTH
PECKING DEPTH
TOTAL HOLE DEPTH

PECK-DRILLING
CONTROL MENU ASKS FOR:
 • SETUP CLEARANCE
 • TOTAL HOLE DEPTH
 • PECKING DEPTH
 • DWELL TIME (SECONDS)
 • FEED RATE

22.   The actual length of the tool must be known to the part
programmer and thus to the machine tool. The tool setting point
defines the end of the tool. In order to avoid the time
consuming requirements of setting the tools to the right lengths,
tools are set to an approximate dimension or size, and then are
corrected to the proper depth after the first cut or by touching
the end of the tool with a probe.

23.   Recalling that process capability refers to the accuracy
and precision of a process, suppose a part is being turned in a
CNC machine. The probe can detect the current size of the part.
The correct depth of cut necessary to bring the part to size can
be calculated by the computer. This depth of cut will account
for variations in tool size and tool wear and deflections in the
machining system.

24.   The preparatory or G functions (also called G words)
precedes the dimension words and prepares the control system for
the information that is to follow in the block of information. G
codes range from G00 to G99.

25.   Computer graphics can be used to verify the NC program. In
addition, a sample part machined out of plastic or machinable wax
can be used to check actual dimensions, although cutting forces
and therefore deflections may be different when cutting metals
than when cutting plastics.

26.   APT stands for Automatically Programmed Tools. APT is a
computer-aided part programming software system which consists of
the input language, the APT processor, an APT postprocessor, and
of course, a computer of sufficient size to handle the APT
program.

1. The X and Y dimensions for hole #2 are +6.7118 and +8.6563
    The X and Y dimensions for hole #3 are +7.9445 and +4.0555

2. $2 \pi r = 360° = 2 \pi 5 = 31.42$ inch
    $\cos(\theta/2) = (5-T)/5 = (5-0.001)/5 = 4.999/5$
        where θ = the span angle
    Therefore, $\theta/2 = 1.1459$ deg and $\theta = 2.292$ deg.
    $AB = (31.42 \times 2.292)/360 = 0.2$ inches

3. $\cos(\theta/2) = (5-T)/5 = (5-0.0001)/5 = .99998$
    $\theta/2 = 0.36$ deg ; $\theta = 0.7247$ deg
    $AB = (31.42 \times 0.7247)/360 = 0.060$ inches

4. $\sin 45° = Y'/6$
    $Y' = 6 \sin 45° = 6 \times 0.707 = 4.24$ inches
    $Y = 10 + 4.24 = 14.24$ inches
    $X = 12 + 4.24 = 16.24$ inches

5. $18 = 2^4 + 2^1$

    $10 = 2^3 + 2^1$

6.
PT	X	Y
1	0	-.5
2	12.5	-.5
3	12.5	12.5
4	-.5	12.5
5	-.5	-.5
6	0	-.5

7
PT	X	Y
1	0.0	-0.5
5	-0.5	-0.5
4	-0.5	12.5
3	12.5	12.5
2	12.5	-0.5
1	0.0	-0.5

8. $\cos 45° = -Y/25$ ipm
    $-Y = 17.68$ ipm
    $-X = 17.68$ ipm

CASE STUDY - CHAPTER 29
Break-Even Point Analysis of a Lathe Part

[NOTE: This case study was developed from real factory data for a real part. Therefore the findings represent the real situation in the job shop where the machines are often employed. In this environment, you assume that each machine has one operator doing the job.]

1). The fixed costs, which do not change with make quantity, are: engineering, tooling and setup. The variable costs are the run cost and the material cost (which was not listed).
2). In order to find the run cost, one has to compute the CT from the equations found in the chapter and then add to that the time needed for part loading/unloading, tool changing and adjusting, inspection, and so forth during each cycle. See Figure 1-6 and references on cost estimating.
3). The nonmachining portions can be estimated from other similar jobs done on these machines, or one can use techniques like MTM. One cannot use time study, since these jobs are not yet setup and running.
4). This time estimate is multiplied by the labor cost per hour, which can include a factor for factory overhead.
5). and 6). These are the plots shown.
7). The breakeven quantities are shown on the plots. The turret lathe is never an economical alternative, being technologically replaced by the NC lathe. We observe that the turret lathe had a very narrow region over which it was economical if one ignored the NC lathe curve. In addition, we observe that none of these processes display an economic minimum, but rather there are regions in which one process is economically preferred. As the build quantities increase and the processes become more automated, the cost per unit continues to decrease, approaching the variable cost per unit as a limit.
8). The turret lathe is never an option. However, if the NC lathe is removed from the solution, the turret lathe has break-even quantities at about 50 units and 75 units (a very narrow range). Turret lathes are being used far less in production operations due to the greater flexibility, capability, and productivity of numerical control lathes.

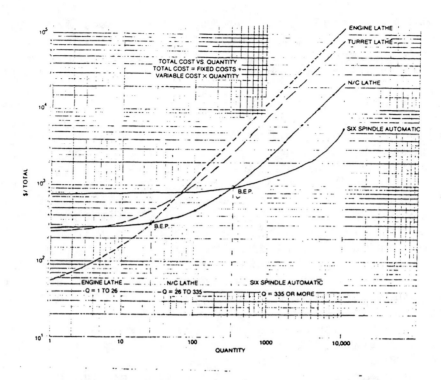

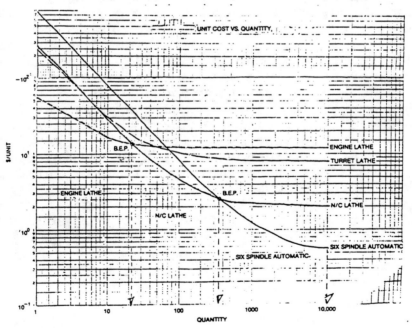

Chapter 30

THREAD MANUFACTURING

1.   The major diameter is the over-all, outside diameter of the thread. The pitch diameter is a smaller, theoretical diameter upon which all the design elements of a thread are based.

2.   The pitch and lead are the same for a single-pitch thread.

3.   The helix angle is the angle between the slope of the screw thread and a line perpendicular to the axis of the screw.

4.   Pipe threads are made on a taper so that as the threaded joint is tightened it will form a liquid-tight joint.

5.   The basic methods for making external threads are: machining (grinding), forming, and casting. In plastics, threads can be molded.

6.   1/4"-20 UNC-3A designated an external thread of the Unified, or American, form, 1/4" nominal diameter, having 20 threads per inch, and a Class 3 fit.

7.   M20 x 2.5-6g6g designates a metric thread; the nominal size is 20 mm; the pitch is 2.5 mm; #6 tolerance grade and "g" tolerance position on the crest diameter. The x means "by".

8.   Fine-series threads are being used less because of the wide availability and use of self-locking plastic inserts on fasteners and special locking coatings.

9.   Pitch is controlled by controlling the longitudinal motion of the lathe carriage relative to the rotation of the spindle, by means of the lead screw and clamp nut. Comment on threading on a lathe: Cutting threads on a lathe is a slow and expensive process. The design should specify standard threads which can be made by the most economical process whenever possible. Can thread rolling be used? If machined threads are needed and if the threads are of standard diameter, they can be cut with a die. Dies come in standard sizes. Nonstandard size threads would require operator controlled functions and great time delay in the cycle to make the threads. This is typically how they are made on the engine lathe, but engine lathe work is only for very small lots. The use of a die allows the turret lathe operations to be performed rapidly, without adjustment. An NC lathe can do threads quickly and repeatably.

10.   The threading dial assures that the cutting tool will exactly "track" in the previous thread groove during successive cuts.

11.   Figures 23-10 (not labeled) & 23-11 show threading dials.

12. The lead is built into the cutting die. It twists itself on to the shaft just like a nut.

13. The purpose of a self-opening die head is to permit the die head to be withdrawn linearly from the completed thread without having to be unscrewed from it.

14. The shape of a taper tap aids in properly aligning the tap in the hole. It is much more difficult to align a plug tap properly if it is not preceded by a taper tap.

15. If full threads are specified to the bottom of a dead-end hole, it is necessary to follow the usual plug tap with a bottoming tap, which must be used with care to avoid breaking off the tap in the hole.

16. A fluteless tap produces threads by plastic flow of the material, requiring a ductile material. Gray cast iron is brittle, and therefore does not plastically flow. Threads in gray cast iron must be machined.

17. If possible, have the hole drilled deeper than actually needed so that it can be threaded to the desired depth without having to use a bottoming tap.

18. A spiral-point tap projects the chips ahead of the tap, thereby avoiding chips from becoming entangled in the cutting tap. (Another reason to drill the hole deeper than the threaded portion.)

19. No, a fluteless tap forms threads progressively, thus requiring several partially formed threads ahead of the fully formed threads, therefore it can not be used to thread a dead-end hole to the bottom.

20. Yes, it is not only desirable but necessary in most materials that the cutting fluid be a good lubricant. There will be large friction forces between the teeth of the tap and the tapped hole as the tap progresses. A lubricant will also reduce the friction between the chips and the work material and the tap.

21. Threads are milled (machined) using form cutters - either single or multiple-form cutters are used. Because the cutter has multiple teeth, the thread can be fully machined in one pass of the cutter past the rotating workpiece. So this process is faster than thread turning, which uses a single point tool.

22. By grinding, threads can be made on hardened materials and the threads will be more precise (less variability) and have a better surface finish.

23. Thread rolling is much faster than any of the machining processes and the properties of the threads are improved --stronger and smoother. The materials to be thread rolled must

be ductile and full-form threads cannot be obtained by rolling. Also, the threads will not have any sharp radius. The surface of the bar must have a good surface finish.

24. Examine the thread. Machined surfaces are very different from rolled surfaces.

25. Flutes are the grooves behind the cutting edges which capture and carry the chips. A tap without flutes produces no chips and forms threads by cold forming.

PROBLEMS FOR CHAPTER 30

1. For 30 fpm, the RPM will be N = (12 x 30)/(3.14 x 0.75) = 152.8 rpm. The cutting time CT = (2 + 0.75)/(0.1 x 152.8) = 0.179 min., where 10 thread per inch = 0.1 inches/thread or 0.1 inches per revolution and 0.75 inches is the allowance for overtravel to insure that full threads are cut.

2. The recommendation is based on the favorable cutting time of the chipless tap over the normal tapping process. The engine blocks may be made out of cast iron, in which case fluteless tapping will not work because cast iron is a brittle material. In addition, tapping deep, dead-end holes with fluteless taps is a difficult process. If these are aluminum engine blocks, then the suggestion should be given serious consideration. P.S. Do not forget to ask the operator and the foreman in the area what they would recommend, as they are going to have to implement any suggestions you make.

CASE STUDY FOR CHAPTER 30
Vented Cap Screws

The redesigned part would need a hole 0.024 inches in diameter, 1/2 inch deep. This is a hole diameter to hole depth ratio of 1 to 20, clearly not a conventional drilling process.

The hole or the slot, as designed, could be made by EDM or ultrasonic machining, or perhaps laser machining, depending upon the available equipment. If none of this equipment is available, ask the designer if the slot needs to be of uniform cross section throughout, or only in some region to control the pressure buildup. This is probably the case, and would allow the slot to be redesigned as shown in the sketch below. This design can be slot milled using a one-inch-diameter slot milling cutter.

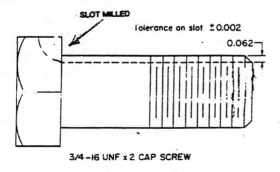

SLOT MILLED
Tolerance on slot ±0.002
0.062
3/4-16 UNF x 2 CAP SCREW

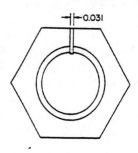

0.031

Chapter 31

GEAR MANUFACTURING

1.   The relative angular velocities of gears are based on the pitch diameter, which is less than the outside diameter.

2.   With the involute tooth form, there is only rolling contact between the gear tooth surfaces, thus eliminating sliding friction.

3.   The diametral pitch of a gear is the ratio of the number of teeth to the pitch diameter, or is the number of teeth per inch of pitch diameter.

4.   The module and the pitch diameter are the same.

5.   See Figures 31-1 and 31-2.

6.   (1) The actual tooth profile must coincide with the theoretical profile; (2) tooth spacing must be uniform and correct; (3) the actual pitch circle must be coincident with the theoretical circle and be concentric with the axis of rotation; (4) the face and flank surfaces must be smooth and adequately hard; (5) the shafts and bearings must assure that center-to-center distances are maintained under load. Notice that most of these requirements are determined solely by manufacturing of the gears.

7.   The tooth engagement of helical gears is gradual, and more teeth are in contact at a given time. This tends to provide smoother and quieter operation. (Do you think that helical gears are used in car transmissions?)

8.   Helical gears cause side thrust and are more difficult to manufacture than straight tooth gears.

9.   The gear would have 24 teeth, and a #5 cutter would be used. See Table 31-2.

10.   A hob has to extend past the point being cut on the gear teeth. On herringbone gears this would cause the hob to extend, and cut into the teeth beyond the centerline of the gear.

11.   Full-herringbone gears can be cut only on a Sykes gear generating machine.

12.   A clearance groove is machined around the center of the gear to provide clearance for the hob, or two helical gears, having opposite helix angles, may be machined separately and joined together.

13.   A different, and very expensive, broach has to be made for each size and type of gear.

14.   A crown gear will mate properly with any bevel gear having the same diametral pitch and tooth form.

15.   Three basic processes for machining gears are: form cutting, generating, and template machining.

16.   The Fellows gear shaper uses generation, meaning the tooth profile is made in progressive passes. See Figure 31-10.

17.   No. The table cannot be rotated about the vertical axis to permit cutting at the required helix angle.

18.   The feed screw of the milling machine table is geared to the dividing head so as to cause the spindle of the dividing head, which holds and rotates the gear blank, to rotate in relationship to longitudinal movement of the table.

19.   The cutters simulate the teeth on a crown gear, and the gear blank is geared to the cradle, which simulates the crown gear.

20.   A hob has almost continuous cutting action, there are multiple teeth, and the action does not have to be stopped to index the gear blank.

21.   The tooth profiles are produced by successive cuts of the cutter past a slowly rotating workpiece.

22.   Cold-roll forming is a very rapid process, the faces of the resulting teeth are very smooth and somewhat hardened, and the gear may be stronger.

23.   Individual gear blanks could be cold-roll formed. One could also extrude the gear profile and then saw gear blanks out of the extrusion to the desired thickness. One might be able to extrude the hole in the center of the gear, if needed.

24.   For only three gears, form cutting the three in one length on a milling machine and cutting off in a lathe to form the separate gears would be attractive.

25.   Cold-roll forming requires a ductile material; gray cast iron is not ductile.

26.   Shaving cannot be used on hardened gears.

27.   Cold-roll forming produces work hardening and thus provides a better wearing surface on the face of the teeth.

28.   Cast iron is soft, and the lapping abrasive would become embedded in the gears, resulting in them not being lapped but rather the teeth would become laps.

29.   Gear inspection checks: hardness; tooth thickness, spacing and depth; tooth profile; surface roughness; and noise.

30.   Gear finishing is accomplished by gear shaving, roll finishing, grinding (good for hardened gears) and lapping for final finishing.

Problems for Chapter 31

1.   RPM of hob = (27.4 x 1000)/(76.2 x 3.14) = 114.5
        RPM of gear blank = 114.5/36 = 3.18

2.   Effective width = 76.2 + (2 x 38) = 153.2 mm
        Time = 153.2/(3.18 x 1.9) = 25.4 minutes

3.   For the HSS cutter milling 4340 steel, selected values for
        V = 100 fpm and f_t = 0.007 inches per tooth would be
            reasonable.

        f_m= nNf_t where N = 12 V/(πD) = 12 x 100/(3.14 x 4) = 95.4
        f_m= 11 x 95.4 x 0.007 = 7.35 ipm

        CT = (L + A)/f_m = (1 + 2.4)/f_m for a 1 inch thick gear
            = 3.4/7.35 = 0.46 min.
        where A = allowance = $2L_A$ = $2\sqrt{t(D-t)}$ = $2\sqrt{.6(3-.6)}$
            for t=0.6, scaled from Figure 31-5

        Each pass takes about 1/2 minute. From Figure 31-5 one can determine that there are 12 teeth which require 12 passes. Assuming that it takes 30 seconds to return the cutter to the start position and index the gear blank 30 degrees, the job would take about 12 minutes. Down milling will be used to get the best finish on the teeth.

4.   No equations were given for gear shaping in the text. The gear shaper can probably do this gear in 1 or 2 minutes, however.

5.   The broach has 10 sets of progressive tooling, so assume the cost of the tooling is $2500. Assume labor costs $10 per hour and machine overhead is 100%. Assume the milling cutter cost is $100 or so. The milling time is 12 minutes per part versus 15 seconds per part for broaching.
        Savings estimate (12-.25)x20/60 = $3.91/part
Assuming all other costs remain the same,
        Additional cost = $2500 - 100 = $2400 for the broach versus
            the milling cutter
        BEQ = 2400/3.91 = 613 parts

        If the company is making over 613 gears and has both machines, the switch is justified.

Comparing to shaping, assume a 1 minute cycle time for shaping.
    Savings estimate (1-.25)x20/60 = $0.25/part
    BEQ = 2400/0.25 = 9600 parts

Broaching would only be preferred to shaping if there were over 9600 gears to be machined. The broach can easily cut that number of gears and the cost of TiN coating is also justified. For example, adding $100 to each broach,
    BEQ = 3400/3.91 = 869 parts

CASE STUDY - CHAPTER 31
The Six-Second Pinion Gear

The carbide tools were designed with the same geometry as the HSS (high-speed-steel) tools. The HSS tools had very large rake and clearance angles and were placed in an unsupported cantilever beam condition. The rapid tool failure had nothing to do with chatter and vibration, nor with tool wear caused by excessive speed. The rapid failure was caused by edge chipping, which created stress concentrations in the brittle carbide. The cantilever design provided no support and the tools were failing in tension. In short, <u>the design of the machine tool had dictated the design of the cutters and the design of the insert teeth</u>. While it may be acceptable to have HSS cutters designed with large rake and clearance angles, carbides need bulk and support at the cutting edges.

(Author's Note): At the time this case took place (@ 1970), TiN coated HSS tools did not exist. The project was "put in mothballs", but was subsequently resurrected by the company when coated HSS tools became available. Using such tools, the machine could be successfully operated. By this time however, another company had invented an extrusion machine that could make helical pinions at a rate of 60 per minute, taking away some of the process rate advantage of the six-second pinion machine.

Chapter 32

NONTRADITIONAL MACHINING PROCESSES

1.   Four types of NTM processes are: chemical, electrochemical, mechanical, and thermal.

2.   The materials used in the future will be harder and stronger, making traditional machining more difficult. A large amount of energy in metal cutting goes into heat which can cause damage to the material of the part. Delicate workpieces are extremely difficult to machine by traditional methods.

3.   The MRR of abrasive waterjet machining is quite low compared to conventional machining. Single point cutting has MRRs of 2 to 20 cu.in./min. AWM has MRRs of 0.01 to 0.001 cu.in./min.

4.   The six basic steps are: (1) preparation of the artwork, (2) photographic production of the negative, (3) application of the emulsion to the workpiece, (4) exposing the workpiece to light passing through the negative, (5) developing the exposed workpiece, and (6) application of the reagent to the workpiece.
         In chemical milling, one often wants to selectively etch certain parts of the material. The material is coated with a coating, called a resist, which, when exposed to certain wavelengths of light, chemically changes and hardens or sets. The unexposed region can be washed away, leaving an exposed region which can be chemically milled or etched. The resist or mask protects the rest of the surface. This technique allows for complex, minute, detailed masks to be developed. The technique is used in microelectronics.

5.   Spraying continuously washes away the debris and keeps the process progressing evenly.

6.   Very thin parts can be blanked.
         Parts with varying thickness can be blanked.
         Many parts can be blanked at the same time.
         No press or expensive die sets are needed.
         No tools to wear out.

7.   The area having the greatest depth is exposed to the reagent first. Next the resist is removed from the area having the next greatest depth, and the work is again exposed to the reagent. This step-by-step procedure is repeated as often as desired.

8.   No. The ratio of the depth to width is too great.

9.   The width of the groove = width of the maskant + (2/3) depth. The width of the mask should therefore be 21 mm.

10.   Yes, but not very satisfactorily. There would be too much
variation in the geometry and metallurgy in and adjacent to the
weld.

11.   Tapered sections are produced by slowly withdrawing the
workpiece vertically from the chemical bath.

12.   Deburring by vaporizing burrs and fins on cast and machined
parts.

13.   ECM is not really related to chemical machining since ECM
is a deplating process that utilizes an electrolytic circuit with
an external power supply.

14.   Hardness is not a factor in ECM and should have no effect
on MRR.

15.   The current density in a material is obviously a function
of the geometry of the part. Small projections, corners, and
things like burrs will have a current density which is higher
than the bulk regions. The MRR is a function of the current
density. The higher the current density, the faster the MRR.
Thus, geometries like burrs preferentially etch faster.

16.   There is no tool wear to speak of in ECM as the tool is
protected cathodically during the process. There may be some
chemical reaction between the tool and the electrolyte when the
power is off, depending upon the materials involved.

17.   ECG is basically an ECM process with minor amounts of
grinding (i.e. cutting, plowing, and rubbing).

18.   ECG is not suitable for grinding ceramics because they are
not conductors. Ultrasonics can be effectively used to machine
ceramics, but the process is quite slow (low MRR).

19.   The MRR in ECM depends mainly upon the current density
which is influenced by the geometry of the tool. For example,
the current density at sharp corners will be greater than flat
surfaces, so corners will cut faster.

20.   The amount of material removed is a function of exposure
time. As the tool advances down into the work, the sides of the
tool would continue to machine the sides of the hole, giving it a
taper (largest at the top) which in this case is not desired.
Thus, the tool is insulated to prevent the passage of current.

21.   The kerf is the slot created by the abrasive jet.

22.   Garnet is diamond (left over particles or scrap or very
poor grade diamonds).

23.   The abrasive jet cuts through the part and must be captured
so that it does not machine surfaces other than the parts.

24.   No, it makes chips just like grinding.

25.   The acceleration is greatest at the ends of the stroke so the forces acting on the grits in the slurry are greatest here as well. Actually, the grits in the slurry are driven by the wave action of the vibrating tool against the workpiece. The tool acts to focus this wave action into the desired regions.

26.   The surface is heated by the sparks to either melt or even vaporize metal. The melted metal is washed away by the dielectric. The sparks cut small, spherical shaped, cups into the surface. The surface is covered with recast (melted and resolidified) metal. Thus, there will always be a hard, brittle, surface layer on EDM part surfaces.

27.   The moving wire electrode can cut straight or angled slots through plates under CNC control. The thin wire allows relatively complex geometries to be cut into dies and stripper plates, for example, with virtually the same program being used to make parts which will later mate with the die set. The principal advantage is that it can produce "saw-like" cuts in hard, delicate materials which would be difficult to bandsaw. There are no tool forces on the wire. A hole can be drilled in the part and the wire passed through the hole.

28.   The effect would not be great. The MRR is controlled by adjusting the amperage (higher amperage, higher MRR) while the surface finish is controlled by the frequency of the spark (higher frequency with amperage constant yields smaller craters and smoother finish).

29.   ECM is probably preferred to EDM in this case because the recast layer produced by EDM may serve as a source for fatigue cracks in the already brittle base material.

30.   Of the four processes, laser beam machining is the easiest to automate into large volume production provided the laser can do the job. Lasers leave recast surfaces like EDM. Both ECM and EDM can be automated but are more oriented toward batch processes. If the parts are small and a large number can be loaded into a machine at the same time, EDM or ECM could be used, with ECM holes having less damage and EDM usually having slightly faster MRRs in most materials. LBM and EBM have very low MRRs which may exclude them from large volume production.

31.   Again, LBM is good for small holes in hard metals and since they are being used for venting, the recast layer should not be a problem. The low MRR rate may make for long machining cycles, so ECM may be preferred. EDM is not preferred for small holes.

32.   Specific power may be high for LBM because the spot area or volume is very small and the coherent beam energy is very large.

33.   The spark in an EDM process literally blasts molten metal
out of the crater.   These globs of material try to assume the
lowest energy state which is spherical.   They cool from the
outside to the inside, so the inside can form a shrink cavity,
making the spheres hollow.   The spheres may also trap gases to
make them hollow.

Chapter 33

GAS FLAME PROCESSES: WELDING, CUTTING, AND STRAIGHTENING

1.   Welding is a process in which two materials, usually metals, are permanently joined together by coalescence, the coalescence resulting from a combination of temperature, pressure, and metallurgical conditions.

2.   The ideal metallurgical bond requires: (1) perfectly smooth, flat or matching surfaces, (2) clean surfaces, free from oxides, absorbed gases, grease, and other contaminants, (3) metals with no internal impurities, and (4) two metals that are both single crystals with identical crystallographic structure and orientation.

3.   Surface roughness is overcome either by force, causing plastic deformation of the asperities, or by melting the two surfaces so that fusion occurs. In solid state welding, contaminated layers are removed by mechanical or chemical cleaning prior to welding or by causing sufficient metal to flow along the interface so that they are squeezed out of the weld. In fusion welding, the contaminants are removed by fluxing agents. Welding in a vacuum also serves to remove contaminants.

4.   When high temperatures are used in welding, the metals may be adversely affected by the surrounding environment. If actual melting occurs, serious modification of the metal may result. The metallurgical structure and properties of the metal can also be adversely affected by the heating and cooling cycle of the weld process.

5.   The combustion of oxygen and acetylene involves a two-stage reaction. In the first stage,
$$C_2H_2 + O_2 \ \text{-->} \ 2\,CO + H_2$$
And in the second stage,
$$2CO + O_2 \ \text{-->} \ 2CO_2$$
$$H_2 + 1/2\,O_2 \ \text{-->} \ H_2O$$

6.   The maximum temperature in an oxyacetylene flame occurs at the end of the inner cone where the first stage of combustion is complete.

7.   The outer zone of an oxyacetylene flame serves to preheat the metal and provides shielding from oxidation, since some of the oxygen from the surrounding air is used in the secondary combustion.

8.   The three types of oxyacetylene flames that can be produced by varying the oxygen-to-acetylene ratio are: a neutral flame, oxidizing flame (with excess oxygen), and a carburizing or reducing flame (with excess fuel).

9.   MAPP, while providing a slightly lower temperature flame, is more dense than acetylene, providing more energy for a given volume. It can be stored safely in ordinary pressure tanks.

10.   The tip size of the torch can be varied to control the shape of the inner cone, the flow rate of the gases, and the size of the material that can be welded. Larger tips permit greater flow of gases, resulting in greater heat input without requiring higher gas velocities that might blow the molten metal from the weld puddle. Thicker metal requires larger tips.

11.   Metals can be thermally cut by oxygen cutting, arc cutting, laser beam cutting and electron beam cutting.

12.   When torch cutting nonferrous metals, the metal is merely melted by the flame and blown away to form a gap, or kerf. When ferrous material is being cut, it is heated to a sufficient temperature that the iron will then oxidize (burn) rapidly in the stream of oxygen that flows from the torch. Thus, the oxyfuel flame first raises the temperature of the metal and then the oxygen content is raised to continue the cutting, the iron oxide being expelled from the cut by the gas pressure of the torch.

13.   When ferrous metal emerges from a continuous casting operation, its temperature will already be above the necessary 2200°F, so only a supply of oxygen through a small pipe is necessary to start and maintain cutting. This is known as oxygen lance cutting (LOC).

14.   The tip arrangement in an oxyacetylene cutting torch is different from that of an oxyacetylene welding torch. The cutting torch contains a circular array of holes through which the oxygen-acetylene mixture is supplied for the heating flame. A larger hole in the center supplies a stream of oxygen to promote the rapid oxidation and blow the formed oxides from the cut.

15.   Cutting torches can be mechanically manipulated by a number of means, including driven carriages (as for straight cuts), template-tracers, CNC machines, and industrial robots.

16.   In order to cut underwater, a supply of compressed air must be added to the torch to provide the secondary oxygen for the oxyacetylene flame and keep the water away from the zone where the burning of the metal occurs.

17.   If a curved plate is to be straightened by flame straightening, the heat should be applied to the longer surface of the arc. The hot metal will be upset and will contract upon cooling, reducing the length of that region.

18.   Flame straightening cannot be used on thin materials because the metal adjacent to the heated area must have sufficient rigidity to resist transferring the buckle from one area to another.

Chapter 34

ARC PROCESSES: WELDING AND CUTTING

1.   Early attempts to develop arc welding were plagued by instability of the arc, requiring great amount of skill to maintain it, and contamination and oxidation of the weld metal resulted from atmospheric exposure at such high temperatures. In addition, the metallurgical effects of such a process were not well understood.

2.   The three types of current and circuit conditions used in arc welding are: alternating current, straight polarity direct current (workpiece is positive), and reversed polarity direct current (workpiece is negative).

3.   In the consumable electrode processes, the electrode melts to supply the needed filler metal to fill the voids in the joint. With a nonconsumable electrode, such as tungsten, a separate metal wire must be used to supply the filler metal.

4.   The three modes of metal transfer that can occur in arc welding are: globular, spray, and short circuit. They are illustrated in Figure 34-2.

5.   Arc welding processes require the specification of: welding voltage, welding current, arc polarity, arc length, welding speed, arc atmosphere, electrode or filler material, and flux.

6.   Electrode coatings can play a number of roles, among them: (1) provide a protective atmosphere, (2) stabilize the arc, (3) act as a flux to remove impurities from the molten metal, (4) provide a protective slag to accumulate impurities, prevent oxidation, and slow the cooling of the weld metal, (5) reduce weld metal spatter and increase the efficiency of deposition, (6) add alloy elements, (7) affect arc penetration, (8) influence the shape of the weld bead, and (9) add additional filler metal.

7.   Coated electrodes are classified by the tensile strength of the deposited weld metal, the welding position in which they may be used, the type of current and polarity (if direct current), and the type of covering. A four or five-digit system of designation is used, as presented in Figure 34-3.

8.   Shielded metal arc electrodes are baked just prior to welding to remove all moisture from the coating, a source of hydrogen in the welds.

9.   The slag coating in a shielded metal arc weld serves to entrap impurities that float to the surface, protect the cooling metal from oxidation, and slow down the cooling rate of the weld metal to prevent hardening.

10.   Electrodes having iron powder in the coating significantly increase the amount of metal that can be deposited with a given-size electrode wire and current.

11.   Continuous shielded metal arc welding faces the problem of providing electrical contact (through the coating) to the center filler-metal wire. Electrode length is limited because the current must be applied near the arc to prevent the electrode from overheating and ruining the coating. Thus, while some continuous arc welding processes have been developed, most shielded metal arc welding is performed with finite length stick electrodes.

12.   In the gas tungsten arc process, inert gas flows to form an inert shield around the arc and pool of molten metal. Argon, helium, or a mixture of the two are commonly used in this process.

13.   Higher deposition rates can be achieved in gas tungsten arc welding by electrically preheating the filler wire or oscillating the filler wire from side to side when making the weld pass.

14.   Gas tungsten arc welding produces very clean welds, and no special cleaning or slag removal is required because no flux is used. All alloys can be welded by this method.

15. Gas tungsten arc spot welding can be used to produce spot welds between two pieces of metal without having access to both sides of the joint. Because of this feature, it can be used to join sheet metal to a heavier framework.

16.   Gas metal arc welding is faster and more economical than shielded metal arc welding because it eliminates the frequent changing of stick electrodes. In addition, there is no slag formed over the weld, the process can be automated, and the welding head is relatively light and compact.

17.   The primary process variables in gas metal arc welding are: type of current, current magnitude, shielding gas, electrode diameter, electrode composition, electrode stickout, welding speed, welding voltage, and arc length.

18.   In the pulsed-arc gas metal arc welding process, a low welding current is first used to create a molten globule on the end of the filler wire. A burst of high current is then applied, which "explodes" the globule and transfers the metal across the arc in the form of a spray.

19.   Because of the reduced heat input and temperatures of the pulsed arc technique: thinner material can be welded, distortion is reduced or eliminated, workpiece discoloration is minimized, heat-sensitive parts can be welded, high-conductivity metals can be joined, electrode life is extended, electrode cooling techniques may not be required, and fine microstructures are

produced in the weld pool. Lower spattering and improved safety are also observed. The high process speed is attractive for productivity, and less power is required to produce a weld than for other processes.

20.   With the flux in the center of an electrode, the electrode is less bulky (since no binder is required to hold it onto the outside) and electrical contact can be maintained directly with the surface of the electrode. Thus, flux-cored arc welding becomes something like continuous shielded metal arc welding.

21.   In submerged arc welding, the flux provides excellent shielding of the molten metal and a sink for impurities. In addition, the unmelted flux provides a thermal blanket to slow down the cooling of the weld area and produce a soft, ductile weld.

22.   Submerged arc welds can be performed at high welding speed, with high deposition rates, deep penetration, and high cleanliness. However, submerged arc welds are generally restricted to flat welds because of the need to form an area of molten slag and keep it in place over the molten weld metal. Extensive flux handling, possible contamination of the flux by moisture, the large volume of slag that must be removed, the high heat inputs that promote large grain sizes, and the slow cooling rates are other negative features of the process. The process is not suitable for high-carbon steels, tool steels, aluminum, magnesium, titanium, lead or zinc.

23.   In bulk welding, iron powder is deposited into the prepared gap (beneath the flux blanket but on top of the backing strip) as a means of increasing the deposition rate. A single pass can deposit as much weld metal as seven or eight conventional submerged arc passes.

24.   In plasma arc welding, it is the flow of hot gases that actually transfers heat to the workpiece and melts the metal.

25.   Plasma arc welding offers greater energy concentration, fast welding speeds, deep penetration, a narrow heat-affected zone, reduced distortion, less demand for filler metal, higher temperatures, and a process that is insensitive to arc length. Nearly all metals and alloys can be welded.

26.   The primary difference between plasma arc welding and plasma arc cutting is the pressure of the gas flowing out of the orifice. At lower pressures, the molten material simply flows into the joint and solidifies to form a weld. At higher pressures, the molten material is expelled form the region and the process becomes plasma cutting.

27.   Stud welding is a special adaptation of arc welding that has been developed to weld fasteners into place.

28.   The ceramic ferrule placed over the stud in stud welding acts to concentrate the arc heat and simultaneously protect the heated metal from the atmosphere. It also serves to confine the molten or plastic metal to the weld area and shapes it around the base of the stud.

29.   In the oxygen arc cutting process, the stream of oxygen is directed onto the hot, incandescent metal. It reacts with the oxidizable metal, liquifies, and is expelled, producing the cut.

30.   If the wire feed rate and other variables of gas metal-arc welding are adjusted so that the electrode penetrates completely through the workpiece, then cutting rather than welding will occur. In gas tungsten-arc cutting, a high-velocity jet of gas now passes through the nozzle to expel the molten metal.

31.   Plasma arc cutting is used to cut high-melting-point metals because this process produces the highest temperature available from any practical source. Virtually any material can be melted and blown from the cut.

32.   Radial impingement of water on the arc was found to provide the necessary constriction of the arc, producing the intense, highly-focused arc needed to make a narrow, controlled cut in plasma arc cutting. Magnetic fields have also been used to constrict the arc.

33.   Compared to the oxyfuel technique, plasma cutting is more economical, more versatile, and much faster. Narrow kerfs and smooth surfaces are produced, and surface oxidation is almost eliminated by the cooling water spray. The size of the heat-affected zone is significantly reduced and heat-related distortion is virtually eliminated.

34.   Heat-affected zones on cut edges of pieces to be welded are of little, if any, concern because this region will be subjected to an additional thermal cycle in the weld that will substantially alter or replace the as-cut structure.

35.   Because of the low rate of heat input, oxyacetylene cutting will produce a rather large heat-affected zone. Arc cutting produces effects similar to arc welding. Plasma arc cutting is so rapid that the heat-affected zone is less than 1/16 inch.

36.   Cutting tends to produce surfaces in residual tension. If subsequent machining removes only a portion of the total surface, the resulting imbalance of stresses may cause warping.

37.   In addition to the effect of residual stresses, Flame- or arc-cut edges are rough and contain geometrical notches that can act as stress raisers and reduce the fatigue performance and toughness of a product. Thus, it is suggested that the cut surface and heat-affected zone should be removed (or at least subjected to a stress relief) on a highly-stressed machine part.

Repair of a Bicycle Frame

The adhesive bonding employed in the original construction was selected because the material properties are heat-sensitive, and the heat of an elevated-temperature joining method would significantly diminish the strength of the tubing material.

1. If the material were cold-drawn tubing, the heat-affected zone created by the weld repair would contain regions of recovery, recovery and recrystallization, and possibly recovery, recrystallization and grain growth. (These phenomena are discussed in Chapter 3, Section 3.16, and the heat affect section of Chapter 39, Section 39.4.) These structures are significantly weaker than the original cold-drawn material, and would be subject to failure by the ductile overload mechanism. While the weld itself was not really defective, the failure occurred as a result of the welding process -- namely the creation of a heat-affected zone that adversely altered the properties of the base material. Therefore, the second failure was indeed the result of the welding repair.

2.   If the tubing had been age hardened material, regions of the heat affected zone would have been hot enough to re-solution treat (and then produce a totally new structure upon subsequent cooldown), while other locations would have been reheated enough to permit overaging. These effects also serve to reduce the strength of the material, and increase the likelihood of ductile overload failure. Once again, the weld itself may have been of high quality, but the welding process was responsible for the alteration of the base material, and the subsequent failure.

3. The repair of these materials would be limited to low temperature methods, such as adhesive bonding or possibly brazing. Both of these methods gain strength by increasing the area of bonding, so the use of some form of large area internal lug or external sleeve would be desirable, as opposed to a simple butt joint repair.

Chapter 35

RESISTANCE WELDING

1.   In resistance welding, pressure is applied initially to hold the workpieces in contact and thereby control the electrical resistance at the interface. After the proper temperature is attained, the pressure is increased to bring about coalescence of the metal.

2.   Because pressure is applied, coalescence occurs at a lower temperature than required for other forms of welding. Many resistance welds are formed without any melting of the base metal.

3.   The total resistance between the electrodes consists of three components: (1) the resistance of the workpieces, (2) the contact resistance between the electrodes and the work, and (3) the resistance between the surfaces to be joined, known as the faying surfaces.

4.   The resistance between the electrodes and the workpiece can be minimized by using electrode materials that are excellent electrical conductors, by controlling the shape and size of the electrodes, and by using proper pressure between the work and the electrodes.

5.   If too little pressure is used, the contact resistance is high and surface burning and pitting of the electrodes can result. If the pressure is too high, molten or softened metal may be squirted or squeezed from between the faying surfaces or the work may be indented by the electrodes.

6.   Ideally, moderate pressure should be applied to hold the workpieces in place and establish proper resistance at the interface prior to and during the passage of the welding current. The pressure should then be increased considerably to complete the coalescence and produce the forged, fine-grain structure.

7.   Resistance spot welding is the simplest and most widely used form of resistance welding.

8.   Spot weld nuggets typically have sizes between 1/16 and 1/2 inch in diameter.

9.   The two basic types of stationary spot welding machines are the rocker-arm machine and the press-type machine. The rocker-arm design is used for light-production work where complex current-pressure cycles are not required. Larger machines and those used at high production rates are generally of the press-type design.

10.   Spot welding guns allow the process to become portable. The welding unit can now be brought to the work, greatly extending the use of spot welding in applications where the work is too large to be positioned on a welding machine (such as automobiles).

11.   A transgun is a spot welding gun with an integral transformer. When accurate positioning is required in an articulated arrangement, such as an industrial robot, transguns may not be attractive because of the added weight of the integral transformer at the end of the arm.

12.   Steel is clearly the most common spot-welded metal.

13.   The practical limit of thicknesses that can be spot-welded by ordinary processes is about 1/8 inch (3 mm), where each piece is the same thickness. When thicknesses vary, a thin piece can be easily welded to another piece that is much thicker than 1/8 inch.

14.   When metals of different thickness or different conductivities are to be welded, they can generally be brought to the proper temperature in a simultaneous manner by using a larger electrode or one with higher conductivity against the thicker or higher-resistance material.

15.   In roll-spot welding, the seam is actually a series of overlapping spot welds, generally produced by two rotating disk electrodes. Continuous seam welding, on the other hand, applies a continuous current through rotating electrodes.

16.   The use of high-frequency current to heat the material for butt welding is advantageous in that it confines the heat to the surfaces to be joined and their immediate surroundings. The lower the frequency, the greater the depth of heating.

17.   Projection welding enables multiple spot-type welds to be made simultaneously, and reduces the problems associated with electrode maintenance.

18.   The number of projections is limited only by the ability of the machine to provide the required current and pressure.

19.   Some of the attractive features of resistance welding processes as techniques for mass production include: (1) they are very rapid, (2) the equipment is semiautomatic or fully automated, (3) they conserve material (no filler metal is required), (4) Skilled operators are not required, (5) Dissimilar metals can be easily joined, and (6) a high degree of reliability and reproducibility can be achieved.

20.   The primary limitations to the use of resistance welding
are: (1) the high initial coat of the equipment, (2) limitations
to the type of joints that can be made, (3) skilled maintenance
personnel are required to service the control equipment, and (4)
some materials require special surface preparation prior to
welding.

21.   Because of the rapid heat inputs, short welding times, and
rapid quenching by both the base metal and the electrodes, the
cooling rates in spot welds can be extremely high. In medium-and
high-carbon steels, martensite can readily form, and a post-weld
heating is generally required to temper the weld.

Chapter 36

OTHER WELDING AND RELATED PROCESSES

1.   The forge welds of a blacksmith were somewhat variable in nature and highly dependent on the skill of the individual because his heat source was somewhat crude, temperatures were uncertain, and it was difficult to maintain metal cleanliness.

2.   Coalescence is produced in cold welding by only the application of pressure. No heating is used, the weld resulting from localized pressures that produce 30 to 50% cold deformation.

3.   By coating portions of one sheet with a material that prevents bonding and then roll bonding with another sheet, products can be made that are bonded only in selected regions. If the no-bond region is then expanded, the expanded regions can form flow channels for fluids.

4.   The heat for friction welding comes from mechanical friction between two abutting pieces of metal that are held together while one rotates and the other is stationary.

5.   Inertia welding differs from friction welding in that the moving piece is now attached to a rotating flywheel. The flywheel is brought to speed, separated from the driving motor, and the rotating assembly is pressed against the stationary member. The kinetic energy is converted to heat during the deceleration.
    In friction welding, the contact is made while the driven piece is connected to the motor, all rotation is stopped, and the pieces are further pressed together.

6.   In the friction and inertia welding processes, surface impurities tend to be displaced radially into a small upset flash that can be removed after welding, if desired.

7.   Friction and inertia welding is restricted to joining round bars or tubes of the same size, or connecting bars or tubes to flat surfaces. In addition, the ends of the workpieces must be cut true and be fairly smooth.

8.   Ultrasonic welding is restricted to the joining of thin materials, such as sheet, foil, and wire, or the attaching of thin sheets to heavier structural materials.

9.   Ultrasonic welding can be used to join a number of metals and dissimilar metal combinations (even metals to nonmetals). Since temperatures are low, the process is an attractive one for heat-sensitive materials. The equipment is simple and reliable, and only moderate operator skill is required. Surface preparation and energy requirements are less than for competing processes.

10   Diffusion welding is a solid state bonding that occurs when properly prepared surfaces are maintained in contact under pressure and time at elevated temperature. Quality of the bond is highly dependent upon surface preparation.

11.   If the interface of an explosive weld is viewed in cross section, it would exhibit a characteristic wavy configuration at the interface. See Figure 36-8.

12.   A thermit weld is quite similar to a metal casting in that molten metal is produced externally and is introduced into a prepared cavity. In the case of the thermit weld, the super-heated metal is produced from the reaction between iron oxide and aluminum, and then flows into a prepared joint providing both heat and filler metal. Runners and risers must be provided, as in a casting, to channel the molten metal and compensate for solidification shrinkage.

13.   In thermit welding, heat comes from the superheated molten metal and slag obtained from the exothermic reaction between a metal oxide and aluminum.

14.   Thermit welding can be used to join thick sections of material, particularly in remote locations where more sophisti-cated welding equipment is not available.

15.   In electroslag welding, heat is derived from the passage of electrical current through a liquid slag. Resistance heating within the slag causes the temperature increase.

16.   In electroslag welding, the molten slag serves to melt the edges of the metal being joined, as well as the fed electrodes supplying the filler metal. In addition, the slag serves to protect and cleanse the molten metal.

17.   Electroslag welding is most commonly used to vertical or circumferential joints because of the need to contain the pool of molten slag. The process is particularly attractive for the joining of thick plates (up to 36-inches thick).

18.   High vacuum is required in the electron beam chamber of an electron beam welding machine because electrons cannot travel well through air. In many operations, the workpiece is also enclosed in the high-vacuum chamber and must be positioned and manipulated in this vacuum.

19. In addition to having to position and manipulate production pieces in a high vacuum, there are size and shape restrictions imposed by the size of the actual vacuum chamber. The high vacuum must be broken and reestablished as pieces are inserted and removed, significantly impairing productivity. If welding is performed on pieces that are outside of the vacuum chamber, high capacity vacuum pumps must be used to compensate for leakage through the electron-emitting orifice. The penetration of the

beam and the depth-to-width ratio of the molten region are considerably reduced as the pressure increases.

20.   High-voltage electron beam welding equipment emits a considerable quantity of harmful X-rays and thus requires extensive shielding and indirect viewing systems for observing the work.

21.   Almost any metal can be welded by the electron beam process. Dissimilar metals can be welded. The weld geometry offers a narrow profile and remarkable penetration. Heat input and distortion are low, and the heat-affected zone is extremely narrow. Welding speeds are high, and no shielding gas, flux, or filler metal is required.

22.   Compared to electron beam welding, laser beam welding: (1) does not require a vacuum environment, (2) generates no X-rays, (3) can employ reflective optics to shape and direct the beam, and (4) does not require physical contact between the workpiece and the welding equipment (the beam can pass through transparent materials).

23.   Laser beams are highly concentrated sources of energy and the resulting welds can be quite small. While the power intensity is quite high, the weld time is extremely small and the total heat input can be quite low. For these reasons, laser beam welds are quite attractive to the electronics industry.

24.   In the laser beam cutting process, a stream of "assist gas" is often used to blow the molten metal through the cut, cool the workpiece, and minimize the heat-affected zone.

25.   Through the use of a fiber-optic cable, laser energy can be piped to the end of a robot arm, eliminating the need to mount and maneuver a heavy, bulky tool that would produce elastic flexing of the components of the robot arm and affect the accuracy of positioning. Cutting and welding can then be performed with the multiple axes of motion of an industrial robot or CNC machine.

26.   Laser spot welding can be performed with access to only one side of the joint. It is a non-contact process, involves no electrodes, and produces no indentations. Weld quality is independent of material resistance, surface resistance, and electrode condition. The total heat input is low, and the heat-affected zone is small.

27.   The flashing action in flash welding must be long enough to provide heat for melting and to lower the strength of the metal to allow for plastic deformation. Sufficient upsetting should occur that all impure metal is squeezed out into the flash and only sound metal remains in the weld.

28.   Only the thermoplastic polymers can be welded, since these
materials can be melted and softened by heat without degradation.
The thermosetting polymers do not soften with heat but tend only
to burn or char.

29.   Thermoplastic polymers can be welded by methods that use
mechanical movement or friction to generate the required heat
(such as ultrasonic welding, friction welding, and vibration
welding), and methods that employ external sources of heat (such
as hot-plate welding, hot-gas welding, and resistive or inductive
implant welding).

30.   Surfacing materials can be deposited by nearly all of the
gas-flame or arc welding methods, including: oxyfuel gas,
shielded metal arc, gas metal arc, gas tungsten arc, submerged
arc, and plasma arc. Laser hardfacing has also been performed.

31.   Several of the thermal spray processes are adaptations of
oxyfuel welding equipment involving some form of material feed.
Electric arcs can be used to melt the material and produce the
molten particles, and plasma spray processes are also quite
common.

32.   Thermal spraying is similar to surfacing and is often
applied for the same reasons. The thermal spray coatings are
usually thinner, and the process is more suited for irregular
surfaces and heat-sensitive substrates.

33.   In metallizing, the bond between the deposited material and
the base metal is a purely mechanical one. To enhance mechanical
interlocking, the surface can be roughened by a variety of
methods, including: grit blasting or the machining of grooves
followed by deformation to roll over the crests or mushroom the
flat upper surfaces.

PROBLEMS FOR CHAPTER 36

1.   This is really an open-ended library-type research
assignment, and the results will vary considerably with the
specific process chosen to investigate.

2.   Thermosetting polymers can be joined by such processes as:
adhesive bonding, threaded fasteners, riveting, and other types
of mechanical joining. Similar restrictions apply to the
elastomers, but threaded fasteners are not as viable. For
ceramic materials, the most common method of joining is adhesive
bonding. Mechanical joining requires the use of large washers or
load distributing devices, and rivets are seldom employed.

Field Repair to a Power Transformer Case

1. The primary restriction here is the need for portability. A process, such as oxyacetylene welding requires only bottled gas, flow regulators and an appropriate torch. These can be scaled and are readily portable. The arc welding methods require a power supply, and an AC, plug-in outlet is not likely to be available. Thus, the electrical capabilities will be limited to those that can be provided by a portable generator. These can be truck-mounted and powered by gasoline motor. The finite-length stick electrodes of the shielded-metal arc process would be quite appropriate for this application because of the wide variety of materials, geometries, and applications encountered in field repair. Gas tungsten arc and gas metal arc are also possibilities. The size and geometry (fillet joint) would not be attractive for the electroslag, submerged arc, or thermit processes, and the equipment required for other alternatives would not be sufficiently portable.

2.- 4. This information can be found by surveying various texts and handbooks. Selection is really a matter of preference, with due consideration to material, the need to weld in both downward and upside-down fillet positions, and the probability of oil contamination and possibly even paint (this is a field repair on an installed item).

Chapter 37

BRAZING AND SOLDERING

1.   Low-temperature production joining methods include: brazing, soldering, adhesive joining, and the use of mechanical fasteners.

2.   Brazing employs a filler metal whose melting point is below that of the metals being joined. It differs from welding in that: (1) the composition of the brazing alloy is significantly different from the base metal, (2) the strength of the brazing metal is substantially lower than the base metal, (3) the base metal is not melted during joining, and (4) bonding requires capillary action.

3.   Since neither of the base metals are melted during the brazing operation, and the bond is formed by introducing a lower melting temperature metal into the gap, the brazing process is attractive for the joining of dissimilar metals, such as ferrous to nonferrous or metals of widely different melting points.

4.   Because brazing introduces a filler metal of different composition from the materials being joined, and the process can be used to join dissimilar metals, brazing can result in the formation of a two-component or three-component galvanic corrosion couple.

5.   If the strength of the brazed joint is to exceed the strength of the metal being brazed, it is necessary to have clean surfaces, proper clearance, good wetting, and good fluidity. The bond strength of a brazed joint is a strong function of the clearance between the parts to be joined. There must be sufficient clearance that the braze metal will wet the joint and flow into it, but further clearance will cause the strength of the braze joint to rapidly drop, approaching that of the braze metal itself.

6.   The clearances necessary for good flow and wetting of the joint are those that are present at the temperature of the brazing process. The effects of thermal expansion should be compensated when specifying the dimensions of the joint components.

7.   Some of the considerations when selecting a brazing alloy include: compatibility with the base materials, brazing temperature restrictions, restrictions due to service or subsequent processing temperatures, the brazing process to be used, the joint design, anticipated service environment, the desired appearance, the desired mechanical properties, the desired physical properties, and the cost.

8.   The most commonly used brazing metals are copper and copper alloys, silver and silver alloys, and aluminum alloys.

9.   Silver solders are silver-copper-zinc alloys that are used in joining steels, copper, brass and nickel. Although they are quite expensive, such a small amount is required that the cost per joint is really quite low.

10.   In brazing, a flux is used to: (1) dissolve oxides that may have formed on the metal surfaces, (2) prevent the formation of new oxides during the heating, and (3) lower the surface tension of the molten brazing metal and promote its flow into the joint.

11.   Since most brazing fluxes are corrosive, the residue should be removed from the work immediately after brazing is completed.

12.   When the braze metal is preloaded into the joints prior to heating, the parts generally must be held together by press fits, rivets, staking, tack welding, or a jig to maintain proper alignment. Care must also be exercised to assure that the filler metal is not drawn away from the intended surface by capillary action of another surface of contact. The flow of filler metal must not be cut off by the absence of required clearances or the presence of entrapped air.

13.   In the torch-brazing process, it is difficult to control the temperature, maintain uniform heating, and meet the cost of the skilled labor required to do the above.

14.   In furnace brazing, reducing atmospheres or a vacuum are frequently used to prevent the formation of surface oxide and possibly reduce any existing oxides and eliminate the need for a brazing flux. If reactive metals must be brazed, a vacuum may be required.

15.   Salt-bath brazing is an attractive process because: (1) the work heats very rapidly because it is in complete contact with the heating medium, (2) the salt bath acts as a protective medium to prevent oxidation, and (3) thin pieces can easily be attached to thicker pieces without danger of overheating because the salt bath maintains a uniform temperature that is less than the melting point of the parent material.

16.   In dip brazing, the entire assemblies are immersed in a bath of molten brazing metal. The braze metal will usually coat the entire workpiece, so such a process is wasteful for all but very small assemblies.

17.   Some of the attractive features of induction brazing are: (1) heating is very rapid, (2) the operation can be made semiautomatic (requiring only semi-skilled labor), (3) the heating can be confined to a localized area (reducing scale, discoloration, and distortion), (4) uniform results are readily obtainable, and (5) by changing coils, a wide variety of work can be performed with a single power supply.

18.   Postbraze operations often include heat treating, cleaning, and inspection.

19.   Fluxless brazing eliminates the operations of flux application and flux removal, both of which can involve significant expense.

20.   Because the strength of a brazed joint is often less than that of the parent metals, the desired strength is often obtained by specifying sufficient joint area (pounds per square inch times square inches = pounds to break the joint).

21.   Braze welding differs from straight brazing in that capillary action is not used to distribute the filler metal. Low melting temperature filler metal is simply deposited into a joint by gravity.

22.   The distinction between soldering and brazing is one of temperature, soldering being a brazing-type operation where the filler metal has a melting point below 840°F (450°C).

23.   Most solders are alloys of lead and tin with the addition of a very small amount of antimony, usually less than 0.5%.

24.   An attractive lead-free solder should possess desirable properties in the areas of melting temperature, wettability, electrical and thermal conductivity, thermal expansion coefficient, mechanical strength, ductility, creep resistance, thermal fatigue resistance, corrosion resistance, manufactur-ability and cost. At present, none of the lead-free solders meet all of these requirements.

25.   Precleaning or surface preparation can be performed by a variety of chemical and mechanical means, including solvent or alkaline degreasers, acid immersion, grit blasting, sanding, wire brushing, and other means of mechanical abrasion. Soldering fluxes further clean and strip away the surface oxides.

26.   Any of the heating methods used for brazing can be used for soldering, but most soldering is still done with electric soldering irons or small torches. For the low-melting-point solders, infrared heaters can be used.

26.   Soldering should not be employed where appreciable strength is desired because soldered joints will seldom develop shear strengths in excess of 250 psi. Butt joints and designs where peeling action is possible should be avoided.

PROBLEMS FOR CHAPTER 37

1.   If the joint clearance is too small, the filler metal may
be unable to flow into the gap and the flux material may be
unable to escape. If the gap is too great, capillary action may
be insufficient to draw the metal into the joint and hold it in
place during solidification.

       Figure 37-1 presents one study of joint strength versus
joint clearance. An optimum value is present, with diminished
properties for either greater or lesser clearance.

CASE STUDY - CHAPTER 37
The Industrial Disposal Impeller

1.   The problem here is one of classical galvanic corrosion.
By using a copper-based brazing alloy in contact with the steel
plate, two dissimilar metals are in electrical contact in the
presence of an electrolyte (water and refuse). The steel
adjacent to the copper becomes anodic and undergoes accelerated
corrosion, causing the support to the carbide pieces to weaken
and ultimately fail. These large pieces of loose carbide then
led to rapid destruction of the unit.

2.   Prevention of this failure requires one to address the
problem of galvanic corrosion. Since it would be difficult to
change the operating environment, the most attractive approach
would involve either breaking the electrical circuit or avoiding
the use of dissimilar metals. Probably the easiest solution
would be to bond the carbide pieces to the steel with a
nonconductive industrial adhesive. This would both remove the
dissimilar metal from the system and prevent the formation of the
corrosive galvanic currents.

Chapter 38

ADHESIVE BONDING AND MECHANICAL FASTENERS

1.   The ideal adhesive bonds to any material, needs no surface preparation, cures rapidly, and maintains a high bond strength under all operating conditions.

2.   Structural adhesives are bonding materials that can be stressed to a high percentage of their maximum load for extended periods of time without failure.

3.   Curing of the structural adhesives can be performed by the use of heat, radiation or light (photoinitiation), moisture, activators, catalysts, multiple-component reactions, or combinations thereof.

4.   After curing at room temperature, high-strength epoxies can develop shear strengths as high as 5000 to 10,000 psi.

5.   Possible additives to epoxy adhesives include: accelerators (to speed the cure rate), plasticizers (to add flexibility, peel resistance or impact resistance), and fillers (to add bulk and reduce cost).

6.   Trace amounts of moisture on the surfaces promote the curing of cyanoacrylates. The anaerobic adhesives remain liquid when exposed to air. However, when confined to small spaces and shut off from oxygen, as in a joint to be bonded, the polymer becomes unstable. In the presence of iron or copper, it polymerizes into a bonding-type resin.

7.   The acrylic adhesives offer good strength, toughness and versatility and can be used to bond a variety of materials. In most cases, a curing agent is applied to one surface and the adhesive is applied to the other. The pretreated parts can then be stored for weeks without damage. Upon assembly, the components react to produce a strong thermoset bond at room temperature. The result is a high-strength bond with room-temperature curing and a no-mix application system.

8.   The silicone adhesives form low-streength joints, but can withstand considerable expansion or contraction. Flexibility and sealing ability are other attractive properties. Numerous materials can be bonded, and the bonds resist moisture, hot water, oxidation, and weathering, and retain their flexibility at low temperatures.

9.   Conductive adhesives can be produced by incorporating selected fillers, such as silver, copper or aluminum flakes or powder. Certain ceramic oxides can provide thermal conductivity.

10.   Temperature considerations relating to the selection of adhesives relate to both the temperature required for the cure and the temperatures likely to be encountered in service. Consideration should be given to the highest temperature, lowest temperature, rates of temperature change, frequency of change, duration of exposure to extremes, the properties required at the various conditions, and the differential expansions or contractions.

11.   Environmental conditions that might reduce the performance or lifetime of a structural adhesive include: exposure to solvents, water, or humidity, fuels or oils, light, ultraviolet radiation, acid solutions, or general weathering.

12.   The stress state in a bonded joint can be tension, shear, cleavage, or peel, as shown in Figure 38-1. Since most adhesives are much weaker in peel and cleavage, joints should be either shear or tension. Looking further, the shear strengths are greater than the tensile strengths, so the best adhesive joint would be one in which the stress state is pure shear.

13.   Surface preparation procedures vary widely, but frequently employ some form of cleaning to remove contaminants and grease. Chemical etching, steam cleaning, or abrasive techniques may be used to further enhance wetting and bonding.

14.   Almost all materials or combinations of materials can be joined by adhesive bonding. The low curing temperatures permit heat sensitive materials and thin or delicate materials to be joined. The resulting bond can tolerate the thermal stresses of differential expansion or contraction.

15.   Most industrial adhesives are not stable at temperatures above 350°F. Oxidation reactions are accelerated, thermoplastics soften and melt, and thermosets decompose.

16.   Adhesives bond the entire joint area. Force equals strength times area. By providing large contact areas, the relatively low strength structural adhesives can be used to produce joints with load-bearing abilities comparable to most alternative methods of joining or attachment.

17.   Adhesives are inexpensive and generally weigh less than the fasteners needed to produce a joint of comparable strength. In addition, the adhesive can also serve to provide thermal and electrical insulation; act as a damper to noise, shock and vibration; and provide protection against galvanic corrosion when dissimilar metals are joined. The adhesive provides both a joint and a seal against moisture, gases, and fluids.

18.   From a manufacturing viewpoint, joint formation does not require the flow of material, as with brazing and soldering, but the adhesive is applied directly to the surfaces. The adhesives can be applied quickly, and useful strengths are achieved in a

short period of time. Surface preparation may be reduced, since bonding can often occur with a oxide film on the surface. Rough surfaces are an asset; tolerances are less critical; and no prior holes have to be made. In addition, the process lends itself to robotics and automation.

19.   Since most adhesives are not stable above 350°F, the structural adhesives would not be attractive for applications that involve exposure to elevated temperatures. At low temperatures, some of the adhesives become brittle.

20.   Selection of a specific fastener or fastening method depends primarily upon the materials to be joined, the function of the joint, strength and reliability requirements, weight limitations, dimensions of the components, and environmental considerations. Other considerations include cost, installation equipment and accessibility, appearance, and the need or desirability for disassembly.

21.   If there is a need to disassemble and reassemble a product, threaded fasteners or other styles that can be removed quickly and easily should be specified.

22.   A press fit differs from a shrink or expansion fit in that mechanical force produces the assembly, not differential temperatures and thermal expansions and contractions. Both involve a strong interference fit to produce a high-strength mechanical joint.

23.   With discrete fasteners, the stress distribution is extremely nonuniform, with much of the load concentrated on the fasteners. With adhesive joints, the load is distributed uniformly over the entire area of the joint.

24.   The most common causes for the failure of mechanically fastened joints relate to joint preparation and fastener installation. Hole manufacture and alignment, installation with too much or too little preload, misalignment of surfaces, insufficient area under load-bearing heads, and vibrations that can lead to further loosening of the joint (and fastener fatigue) are all areas of concern.

25.   The use of standard fasteners would enable ready access at reasonably attractive cost. Nonstandard fasteners require scheduled production, possible delays and additional expense. By minimizing the variety of fasteners within a given product, there is a reduced likelihood of mixup or exchange of pieces during a disassembly and reassembly, or even within the initial production and assembly line. Moreover, inventory costs could be reduced, and by using larger quantities of a given fastener, a reduced price might be available.

PROBLEMS FOR CHAPTER 38

1.   When iron (steel) is galvanically coupled with passive aluminum, iron becomes the anode and undergoes preferential corrosion. With moisture being the electrolyte that completes the electrical circuit, we have a corrosion cell with very small corroding anodes (the heads of the iron nails) and large cathodic surfaces (the aluminum siding). The heads of the nails will rapidly corrode and the siding will eventually separate from the house. Aluminum siding should be installed with aluminum nails!

2.   Hole preparation would be a major area of concern, because we must now produce holes in a fiber-reinforced material. Mechanical means will tend to produce frayed surfaces. Thermal means may damage the fibers and matrix.

      Joint design is also a concern. While the composite material may offer attractive strength properties, these properties may not be present around a fastener where the continuity of the fibers has been disrupted. Screws and similar threaded fasteners will be limited by the strength of the polymeric matrix. Compression fasteners, such as bolts and rivets, may require the use of large washers to spread the load over a larger area. A variety of service-type failures could be considered.

CASE STUDY - CHAPTER 38
Golf Club Heads

1.   Since the club head is a martensitic stainless steel, it achieves its strength by a quench and temper heat treatment. Subsequent exposure to temperatures in excess of the tempering temperature will result in a further loss of strength and hardness. In addition, exposure to temperatures near 1000°F will enable the atomically-dispersed carbon and chromium atoms to diffuse and unite to form chromium carbides. The depletion of chromium will leave the adjacent regions with less than 12% chromium free to react with oxygen to form the protective (corrosion-resistant) oxide. The stainless steel is no longer "stainless" and will be subject to red rust. For these reasons, coupled with possible warping of the thin insert, the joining method is limited to low temperature methods. While brazing or, more preferably, soldering might be possibilities, these methods provide metallic joint, and the electrical conductivity coupled with the presence of two or more dissimilar metals creates a galvanic corrosion cell in a product that may be exposed to humidity and moisture as they are stored in car trunks and other locations. Rivets, screws and other fasteners are possible, but the joining becomes localized, and the possibility of gaps and related dampening is a real one. Among the low-temperature methods, it appears that some form of adhesive bonding would be the most preferred means of assembling the components.

2.   The same problems with the martensitic stainless steel

restrict the temperature of the joint. Most of the above methods continue as options, with brazing or soldering being eliminated because of the polymeric shaft, and shrink or press fits becoming additional alternatives. If the shaft were metallic, brazing or soldering reenter the picture. If the shaft is sufficiently solid, some form of hole and rivet is a possibility.

3.   If the same procedure is to be applied to both joints, one between dissimilar metals, and the other between stainless steel and fiber-reinforced epoxy, then some form of adhesive would appear to be preferred. NOTE: It may also be desirable to consider alternative means of creating the composite club face, such as flame-spray deposition of the carbide-containing surface -- which would eliminate the need to bond two dissimilar metals.

Chapter 39

MANUFACTURING CONCERNS IN WELDING AND JOINING

1.   Process selection is a complex procedure because of the large number of available processes, the variety of possible joint configurations, and the number of parameters that must be specified for each operation.

2.   Common weld defects include: cracks in a variety of forms, cavities (both gas and shrinkage), inclusions (slag, flux, and oxides), incomplete fusion between the weld and base metals, incomplete penetration (insufficient weld depth), unacceptable weld shape or contour, arc strikes, spatter, undesirable metallurgical changes (aging, grain growth, or transformations), and excessive distortion.

3.   The four basic types of fusion welds are bead welds, groove welds, fillet welds, and plug welds, as illustrated in Figure 39-1.

4.   Fillet welds are used for tee, lap, and corner joints. These configurations are shown in Figure 39-4.

5.   The cost of making a weld is affected by the required edge preparation, the amount of weld metal that must be deposited, the type of process and equipment that must be used, and the speed and ease with which the welding can be accomplished.

6.   When two pieces are welded together, they become one piece. Cracks in one segment can then cross the weld and continue propagation throughout the structure. Also, one segment constrains the others, so that properties such as fracture resistance and ductility can change appreciably.

7.   The notch-ductility characteristics of metal can change markedly with a change in the size of the piece. While a small piece, such as a Charpy impact specimen, exhibits ductile behavior and good energy absorption down to low temperatures, a large structure of the same metal exhibits brittle behavior at higher temperatures. Because of the added constraint of mass, deformation and fracture modes that may absorb energy may be forbidden, resulting in a product with reduced fracture resistance, reduced ductility, and an elevated ductile-to-brittle transition temperature.

8.   Excessive rigidity in a welded structure can restrict the material's ability to redistribute stresses, and thereby avoid failure. Structures and joints should be designed to have some flexibility.

9.   In a fusion weld, a pool of molten metal is created, contained, and solidified within a metal mold formed by the segments being welded. This is actually a casting in a metal mold and has all of the structural and property features of such a casting.

10.   The chemistry of a weld fusion zone may be complex because it is a combination of the filler metal and melted-back metal from the material being welded. See Figure 39-7.

11.   Since the solidified weld will be in an as-cast condition, its properties and characteristics will not be those of the same metal in the wrought state. Therefore, electrode or filler metals are usually not selected on the basis of matching chemistry, but on the basis of having properties in the as-solidified or as-deposited condition that equal or exceed those of the base metal.

12.   Fusion weldments may exhibit all of the problems and defects observed with castings, including: gas porosity, inclusions, blowholes, cracks, and shrinkage. Rapid solidification and cooling may lead to: inability to expel dissolved gases, chemical segregation, grain-size variation, grain shape problems, and orientation effects.

13.   In the heat-affected zone, temperature and its duration vary widely with location. This variation in thermal history produces a variety of microstructures and a range of properties.

14.   Structure and property variations in heat-affected zones can include: phase transformations, grain growth, precipitation (or overaging), embrittlement, and even cracking.

15.   Due to possible changes in structure, the heat-affected zone may no longer possess the desirable properties of the parent metal, and, since it was not molten, it does not have the selected properties of the weld metal. Consequently, it is often the weakest area of the weld in the as-welded condition. Except when there are obvious defects in the weld deposit, most welding failures originate in the heat-affected zone.

16.   Processes with low rates of heat input (slow heating) tend to produce a high total heat content in the metal, slow cooling rates, and a large heat-affected zone.

17.   When attempting to heat-treat products after welding, numerous problems can arise in producing controlled heating and cooling in the often large, complex-shaped structures that are typically produced by welding. Furnaces, quench tanks, and related equipment may not be available to handle the full size of the welded assembly.

18. Pre- and post-heating operations can reduce the variation (and sharpness in the variation) in microstructure. The cooling rate in both the weld deposit and adjacent heat-affected zone is reduced, producing more gradual changes in microstructure.

19. In brazing and soldering, the base and filler metals are usually of radically different chemistry. The elevated temperatures of joining can promote interdiffusion. Intermetallic phases can form at the interface and alter the properties of the joint - usually imparting loss of both strength and ductility.

20. The first type of residual stresses are those that result from the restraint to thermal expansion and contraction offered by the pieces being welded. As the hot weld is being made, the adjacent metal becomes hot and expands. The underlying parent metal remains cool, provides restraint, and may force upsetting of the hot, expanding metal. As the weld area cools and wants to contract, the contraction is again resisted by the strong, parent metal.
     If the base metal segments are restrained from movement perpendicular to the line of the weld, reaction residual stresses (the second type) can be induced. Their magnitude is inversely related to the distance between the weld and the restraint.

21. The residual stresses in the weld region of welded structures is nearly always one of residual tension. If the magnitude of these stresses becomes sufficiently large, cracking of the metal may result.

22. The magnitude of the reaction stresses is an inverse function of the length between the weld joint and the point of fixed constraint.

23. The most apparent result of the thermal stresses induced by welding is a distortion or warping as the material seeks to reduce the imbalance.

24. The amount of distortion in a welded structure can frequently be reduced by: forming the weld with the least amount of weld metal necessary to make the joint; use faster welding speeds to reduce the amount of heating of adjacent metal; use the minimum number of welding passes; permit the base metal segments to have as much freedom of motion as possible; and, weld to the point of greatest freedom (as from the center to the edge). Weld surfaces can be peened to induce offsetting compression.

25. When welded structures are subsequently machined, the material removal frequently unbalances the residual stress equilibrium, and the material distorts to achieve a new balance of forces. In essence, it distorts during machining. Weldments that are to undergo appreciable machining should be given a stress-relief heat treatment prior to the machining operation.

26. The cracking of weldments can be reduced by designing joints to keep restraint to a minimum, and selecting metals and alloys with welding in mind. Thin materials are more resistant to cracking than thicker sections. The size and shape of the weld bead should be properly selected and maintained. Weld profile (penetration depth) can affect cracking. By slowing the cooling of the weld area and inducing plasticity into the material, the tendency to crack can be further reduced. Preheats, postheats, and stress reliefs can be used, along with efforts to remove hydrogen from the weld area.

PROBLEMS FOR CHAPTER 39

1. Some possible corrective measures to eliminate or reduce cracking include: (1) possible use of a lower carbon steel, (2) substitution of a low-hydrogen type electrode, and (3) use of preheating and possibly some post-heating if the carbon content of the steel cannot be reduced.

2.   **(a).** By casting:

Cost of the pattern	= $ 450
Cost of the casting = 1400 x $.60	= $ 840
TOTAL COST	= $1290

By welding

Steel = 800 x $.14	= $ 112
Preparation and setup = 30 x $10	= $ 300
Welding labor = 55 x $9.50	= $ 523
Electrode = 200 x $.17	= $ 34
TOTAL COST	= $ 969

Therefore, welding would be more economical for the production of one part.

(b). $840 + $450/N = $969 ---> N = 3.5 or 4 units

Casting would be cheaper if 4 or more units are to be made.

(c). Steel does not possess the damping characteristics of cast iron, and this is important in machine tools as the base of the machine absorbs and damps out many of the vibrations set up in the machine by the motor and other rapidly rotating members. These vibrations, if not damped out, can sharply reduce the accuracy and precision of the machine tools. Thus, a process capability study should be performed on the new design (steel) to compare it with the existing cast iron structure.

3.   The assessment is not fair, because a subsequent
examination of the riveted ships revealed a number of similar
cracks. These cracks, however, simply traveled to the edge of
the plate and stopped. Welding, on the other hand, produces
monolithic structures. The cracks can cross the welds and
continue into and through adjacent pieces. While the problem was
a material problem, it became far more apparent when welding was
used to produce the large, one-piece assemblies.

       The problem was later identified as a metallurgical one
related to the high ductile-to-brittle transition temperatures of
the steel being welded. Additional knowledge of the phenomena,
coupled with the selection of materials with lower transition
temperatures, has permitted the safe use of welded-hull ships
under most of the temperatures likely to be encountered.

       (NOTE: While not a welded-hull ship, it is this same
ductile-to-brittle transition phenomena that is suspected a
playing a significant role in the sinking of the Titanic.)

CASE STUDY - CHAPTER 39

The Welded Frame

1.   Based on the desire to minimize constraint, one should
resist the natural tendency to fabricate the exterior box and
then insert the interior subsections. Instead, the preferred
sequence would begin with the innermost welds and progress
outward. The initial welds might be 4 and 5 -- then 8 and 9.
Welds 3, 7, 6 and 10 would follow, and then on to the final
assembly at 1 and 2 and 11 and 12.

2.   Various "rules" could be proposed, each designed to reduce
the amount of restraint on the weld or the number of welds that
must be made under restraint. When restrained welds must be
made, efforts could be made to maximize the length of material,
or distance to the restraint. (NOTE: If 3/100-inch elastic
stretching must be provided to compensate for weld shrinkage,
this would require a 3% stretch for a 1-inch segment, 0.3% for a
10-inch segment, and 0.03% for a 100-inch segment.)

Chapter 40

SURFACE TREATMENTS AND FINISHING

1.   Some of the possible objectives of surface modification processes are: clean surfaces and remove surface defects, modify a product's appearance, improve resistance to wear or corrosion, reduce friction or adhesion, and conserve costly materials.

2.   When selecting a surface modification process, one should consider the common factors of time, labor, equipment, and material handling. In addition, consideration should be given to such features as: the size of the part, the shape of the part, the quantity to be processed, the temperatures required for processing, the temperatures encountered during subsequent use, and any dimensional changes that might occur due to the treatment.

3.   Manufactured products frequently contain foreign material on the surfaces. Sand from casting molds and cores often adheres to surfaces. Scale can be produced when metals are processed at elevated temperatures. Oxides can form during storage.

4.   Blasting or other abrasive cleaning operations utilize abrasives such as: sand, steel grit, metal shot, fine glass shot, walnut shells, dry-ice pellets, and even baking soda.

5.   Barrel finishing operations are most effective when large quantities of small parts are to be processed.

6.   In barrel finishing, the rotation of the barrel causes the material to rise until gravity causes the uppermost layer to cascade downward in a "landslide" movement. If the barrel is too full, the relative motion between the work and the abrasive will not adequate. Increasing the speed causes the material to rise higher in the barrel, but if the speed is too high, centrifugal forces cause the parts to adhere to the outside of the barrel, thereby eliminating the cascading action.

7.   In barrel burnishing, no cutting action is desired. The parts are tumbled against themselves, or with media designed to produce a peening or rubbing action. In abrasive barrel finishing, material is removed from the surface by abrasive cutting.

8.   In barrel finishing, most of the finishing occurs when the parts slide down over the media. In vibratory finishing the entire load is in constant agitation, and there is virtually constant relative motion between the work and the media.

9.   Synthetic abrasive media, formed by combining abrasive material and a binder, are manufactured, and have consistent and reproducible size and shape.

10.   The compounds that are used in abrasive finishing operations can perform a variety of functions, including: deburring, burnishing, abrasive cutting, cleaning, descaling, and corrosion inhibition.

11.   The ideal geometry for belt sanding is a flat surface. The process is more difficult with curved surfaces, and is extremely difficult to apply when the geometry includes recesses or interior corners.

12.   Electropolishing is the reverse of electroplating, since material is removed from the surface rather than being deposited.

13.   Alkaline cleaning can remove a variety of soils, including oils, grease, wax, fine particles of metal, and dirt. The actual cleaning occurs through one or more of: (1) saponification, (2) displacement, (3) dispersion or emulsification, and (4) dissolution. These are further explained on page 1091.

14. Solvent cleaning cannot be used to remove insoluble contaminants, such as metal oxides, sand, scale, and the inorganic fluxes used in welding, brazing and soldering.

15.   Environmental issues have made vapor degreasing rather unattractive. The standard solvents have been identified as ozone-depleting compounds, and have been banned from use. Replacement solvents usually lack one or more of the qualities that are desirable for the process.

16.   Acid pickling operations are generally used to remove oxides and dirt that remain on the surface of metals after other processing operations.

17.   Burrs are the small, sometimes flexible projections of material that adhere to the edges of workpieces that are formed by cutting, punching or grinding.

18.   During thermal energy deburring, the parts are loaded into a chamber, which is then filled with a combustible gas mixture. When the gas is ignited, the short-duration wavefront heats the small burrs to extremely high temperatures, while the rest of the part remains cool. The burrs are vaporized, including those in inaccessible or difficult-to-reach locations.

19.   While both coating and cladding are deposition processes, coatings are deposited as a liquid or a gas (or from a liquid or gas medium), while the added material is solid during cladding.

20.   Paints are used for a variety of reasons, including providing protection and decoration, filling or concealing surface irregularities, changing the surface friction, and modifying the light or heat absorption or radiation characteristics.

21.   In a painted surface, the prime coat serves to promote adhesion, fill minor porosity or surface blemishes (leveling), and improve corrosion resistance. The more highly pigmented final coats are designed to provide color and appearance.

22.   In airless spraying, mechanical pressure forces the paint through an orifice under pressure. The resultant velocity is sufficient to produce atomization and propel the particles toward the workpiece.

23.   Industrial robots can mimic the movements of a human painter, while maintaining a uniform separation distance and minimizing waste. Monotonous and repetitious movements can be performed with consistent results, and the human operator is freed from an undesirable working environment.

24.   Electrostatic spraying greatly reduces paint loss and the generation of airborne particles, and provides for more uniform coverage of the workpiece.

25.   In electrostatic spray painting, the workpiece must act as one of the electrodes. Wood and plastic are not electrically conductive and cannot serve as an electrode.

26.   The most common metallic coatings that are applied by hot dipping are: zinc, tin, aluminum, and terne (a lead-tin alloy).

27.   The two most common types of chemical conversion coatings are chromate and phosphate.

28.   Nonconductive materials, such as plastic, can be electroplated provided that they are first coated with an electrically-conductive material. Processes, such as the electroless deposition of nickel can be used.

29.   Hard chrome plate offers Rockwell C hardnesses between 66 and 70, and can be used to build up worn parts, and coat tools and other products that can benefit from the reduced surface friction and good resistance to wear and corrosion.

30.   Some of the process variables in an electroplating cell include: the electrolyte and the concentrations of the various dissolved components, the temperature of the bath, and the electrical voltage and current.

31.   Ordinarily only one type of workpiece is plated at a time, since the details of solutions, immersion times, and current densities are usually changed with changes in workpiece size and shape.

32.   In the electroforming process, the coating is stripped from the substrate and becomes the final product. In electroplating, the coating and substrate remain intact.

33. When anodizing aluminum, if the oxide is not soluble in the anodizing solution, the oxide will grow until the resistance of the oxide prevents the current from flowing. If the oxide is partially soluble, dissolution competes with oxide growth and a porous coating is produced. As the coating thickens, the growth rate decreases until it achieves a steady state where the growth rate is equal to the rate of dissolution.

34. In color anodizing, a porous oxide is first produced, which is then immersed in a dye solution. The dye is then trapped in place by a sealing operation.

35. Electroless plating will provide uniform thickness on complex shapes and it requires much less energy. In addition, metallic platings can be directly applied to nonconducting surfaces.

36. When minute particles are codeposited with the electroless metal, electroless plating can be used to produce composite coatings. Commercial applications have used diamond, silicon carbide, aluminum oxide, and teflon particles dispersed in the metal matrix.

37. Mechanical plating is an adaptation of barrel finishing in which coatings are produced by cold-welding soft, malleable metal powder onto a substrate.

38. Porcelain enamel coatings can be used to impart resistance to corrosion and abrasion, decorative color, electrical insulation, and the ability to function in a high-temperature environment.

39. The term PVD applies to a group of processes in which the material to be deposited is physically carried to the surface of the workpiece. All are carried out in some form of vacuum, and most are line-of-sight, in which the workpiece must be positioned relative to the source. CVD processes deposit material through chemical reactions, and generally require significantly higher temperatures than their PVD counterparts.

40. Vacuum metallized coatings are very thin, and are generally used to provide decorative appearance, produce reflective surfaces, or provide electrical conductivity.

41. Even though the deposit in vacuum metallizing is formed from a vapor, the high vacuum minimizes the likelihood of atomic collisions, and emitted atoms travel in a straight line from the source. Deposition, therefore, requires that the surfaces be in a line-of-sight orientation with respect to the source.

42. Because of the kinetic energy of the impact and the cleanliness of the substrate, the adhesion of sputtered coatings is considerably better than with other deposition techniques.

43.   The chemical vapor deposition process can be used to deposit a variety of metals, as well as ceramic carbides, nitrides, borides, silicides and oxides. Refractory metals, alloys, and refractory compounds can be deposited onto complex-shaped substrates.

44.   Diamond thin films have hardnesses greater than the traditional carbides and nitrides, high chemical stability, low coefficient of sliding friction, optical transparency in the optical and infrared spectrums, high thermal conductivity, and high electrical resistivity.

45.   Ion implantation is a physical, rather than a chemical or thermal process. Any ionizable species can be introduced into ant target material, while maintaining temperatures below 300°F. The amount added can be precisely controlled through beam current and exposure time, and the ultimate amount is determined by equilibrium with sputtering, and not solubility limits. The beam can be positioned by electromagnetic controls, and the depth is a function of acceleration voltage. The dimensions, surface finish, and appearance of the target are unchanged by the process.

46.   In ion-beam mixing, a thin film of metal or compound is first deposited onto the target surfaces, and the energetic ion beam is then used to "mix" the deposited material into the surface of the product.

47.   By using coil-coated sheets, the product is effectively prefinished, and efforts need only be made to protect the surface during handling and fabrication. This is in lieu of the normal approach of fabricate and then finish the products on an individual basis.

48.   Each of the various machining processes produce characteristic surface textures on the workpiece. In addition, they produce characteristic changes in the physical, mechanical and metallurgical properties on or near the newly-created surfaces. Machining induces plastic deformation, and the cut surfaces are generally left with tensile residual stresses, microcracks, and a hardness that is different from the bulk material.

49.   Residual stresses couple with applied stresses to produce a net effect. Products with residual compression on the surface, will have a lower net tension and less likelihood of fracture or fatigue when placed under tensile loading. See Figure 40-17.

1.   Because of the wide variety of finishes desired, a number of processes can be considered. Of particular importance is the need to retain the necessary mechanical properties that were set by the heat treatment. Exposure to high elevated temperatures will overtemper the hook, making it prone to possible unwanted deformation. In addition, the need to maintain sharpness of the points and barbs significantly restricts the thickness of any applied coatings. The shape and the presence of the barbs may well make any mass treatment process rather difficult.

The ideal process would produce a durable, thin, uniform-color coating under low temperature conditions. Among the attractive possibilities are chemical conversion treatments, the various PVD methods, and blackening or coloring treatments.

2.   This is a rather open-ended problem offering a wide variety of base materials, sizes, shapes and requirements. It is designed to expose the student to the wide variety of surface treatment methods that are encountered in everyday products.

The extent to which the problem is treated can vary greatly. For example, part k) - the automobile muffler - offers the option of having to integrate surface treatment and manufacturing process. For example, spot or seam welds are quite difficult if the sheet material has an existing galvanized (zinc) coating. Low-carbon sheet steel, however, can be readily seam welded. Therefore, if the sheet material is pre-galvanized, then assembly might utilize some form of mechanical (such as roll-lock) seam. If uncoated material is selected with the intent of subsequent coating, the student should realize that the presence of inlet and outlet tubes, and internal components or baffles, would preclude the possibility of approaches such as hot-dip galvanizing after assembly.

CASE STUDY - CHAPTER 40
Burrs on Yo-Gi's Collar

1).   Here is a listing of deburring techniques commonly used in industry (as contributed by L.K. Gillespie and J.G. Bralla):

- Barrel tumbling. A large group of parts with burrs are placed in a rotating barrel with small pebble-like media, a fine abrasive powder, and water, and the barrel and its contents are slowly rotated until the burrs wear off, typically 4-12 hr.
- Centrifugal barrel tumbling, similar to barrel tumbling except that the barrel is placed at the end of a rotating arm. The addition of centrifugal force up to 25G to the weight of the parts in the barrel makes the process 25-50 times faster than conventional barrel tumbling.

- Spindle finishing, also similar to barrel tumbling except that the workpiece is fastened to the end of a rotating shaft and then placed in a barrel rotating in the opposite direction. The abrasive media gently wears off the burrs and produces a smooth radius on the edges. Although each part must be handled individually, deburring requires only 1-2 min. per part.
- Vibratory deburring, also similar to barrel tumbling except that parts and media are vibrated rather than rotated.
- Abrasive-jet deburring. A high-velocity stream of small abrasive particles or miniature glass beads is sprayed at the burrs, and a combination of impact, abrasion, and peening actions breaks or wears away the burrs. Deburring takes from 30 sec to 5 min, depending on workpiece size and complexity. This process is essentially a refined sand-blasting process.
- Water-jet deburring. A 0.25-mm-dia (0.010-in.-dia) jet of water at very high velocity cuts burrs and flash from the workpiece. In nonmetals, this process can deflash contours at a rate of 250 mm/s (600 ipm).
- Brush deburring. Motorized rotating brushes abrade burrs from parts at a rate from 10 sec to 5 min per part. At least 50 different types of brushes are in common use.
- Sanding. Belt sanders deburr flat parts and disks, and flap wheels deburr contoured parts, at a rate of 600 stamped parts per hour. Although heavy burrs are removed, the process itself often produces a very small burr.
- Mechanical deburring. A variety of specialized machines are equipped with chamfering tools, knives, or grinding wheels to mechanically remove cut burrs. Automotive gears, for example, can be deburred at a rate of 400 gears per hour.
- Abrasive-flow deburring. Hydraulic cylinders force an abrasive-laden putty-like material over burrs at a rate of 30 parts per hour. Deburring at up to 400 parts per hour is possible with automation. Some dimensional changes occur on surfaces contacting the putty-like material.
- Liquid-hone deburring. A 60-grit abrasive suspended in water is forced over burr-laden edges, removing very fine burrs in a 5-min cycle and producing minimal edge radii.
- Chemical deburring. Buffered acids dissolve burrs from large groups of small parts in 5-30 min, depending on the part. Because acids attack all surfaces, some dimensional changes of the entire part occur.
- Ultrasonic deburring. A combination of buffered acids and a fine abrasive media is ultrasonically agitated to wear and etch minute burrs, such as those produced in honing.
- Electropolish deburring. Reverse electroplating operation uses electrolysis in a mild acid solution to remove burrs from all surfaces, producing excellent surface finish.
- Electrochemical deburring, similar to electropolish deburring except that a salt solution and a shaped electrode are required. Stock is removed only at the edges, although some light etching occurs at other places on the part. Surface residues on the part have to be brushed or wiped off.

Deburring typically requires 2 min per part, without automation.
- Thermal-energy deburring. A high-temperature wavefront -produced by igniting natural gas in a closed container - vaporizes burrs. The short-duration wavefront exposes components to only 95C (200F) while burrs are exposed to as much as 3300C (6000F). Up to 80 parts per hour can be deburred by this process.
- Manual deburring. Workers with special knives, files, scrapers, and other tools cut burrs from parts.

2). The screw machine operation has six steps:
   I. Use form tool to turn the 12.7 mm diameter (The original diameter is 17.46 mm.)
   II. Slot the end. (Rotation must be stopped to form the slot.)
   III. Turn the 16 mm diameter. (The original diameter is 17.46 mm. This operation removes the slotting burrs.)
   IV. Drill 6.4 mm hole. (This operation uses the part of the hole remaining from the previous part as a start.)
   V. Thread the 12.7 mm diameter. (Use a thread-forming tool.)
   VI. Cut off. (Cut to the specified 12.7 mm length.)

Drilling and turning after slotting eliminates most of the burrs. This part takes about 20 seconds, so the machine can make 180 per hour. The total run takes about 138 hours or 17-18 days, assuming one 8-hour shift per day. The part would cost about 10 or 11 cents, exclusive of material to make on the screw machine.

3). The collars could also be made by powder metallurgy. It is likely that powder metallurgy would be somewhat cheaper for the quantity required, but much would depend on local conditions as to competition between suppliers.
The collars could also be made by on cold heading machines followed by thread rolling. Quantities may be a bit too small to justify the dies.
The collars could also be investment cast. The cost here would likely be higher than the other alternatives that can take advantage of the geometric symmetries.
The collars cannot be die cast because the melting temperature of the material (AISI 304 stainless steel) is too high.

Chapter 41

MANUFACTURING SYSTEMS AND AUTOMATION

1.   The elimination of costly setup devices or setup time makes it as economical to produce small lots as it does large lots, as the cost per unit is now controlled by the variable cost (direct labor, direct material). Small lot production enhances quality, on-time production, and makes the manufacturing system more flexible.

2.   The parts are passed from machine to machine in a transfer line by means of a conveyor system. Once the parts are loaded on to the conveyor (the transfer mechanism), they stay on the conveyor until they are unloaded, having passed through many stations. The load and unloading station is at the start or end of the line and does not delay the line. Transfer lines are setup for a specific product, and run that product and no other. Therefore there is not setup time between products or parts once the machine is setup, debugged, and operating.

3.   The four classic manufacturing systems are the job shop, the flow shop, the project shop, and continuous flow.

4.   Universities, barber shops, hospitals, and sitdown restaurants are examples of service job shops. A TV show, a stage play, and a movie production are examples of the project shop. A cafeteria would be an example of a flow type operation. The series of health inspections, tests, etc. a service man goes though at the time of induction are organized as a flow shop. Since most service operations are to serve people, which arrive as discrete objects, there are no examples of continuous flow shops that we can think of.

5.   The new form of manufacturing system is called lean production. This new manufacturing system is composed of linked-cells. There are two kinds of unit cells; manufacturing cells and assembly cells. Either of these can be manned or unmanned (automated). Cells operate on a family or group of parts and the parts move past the machines one at a time with no back flow. The workers in the cell are multifunctional and thus perform such duties as quality control, setup reduction and machine tool maintenance in addition to part loading/unloading, deburring, transportation, and inspection. Because cells function like subassembly and assembly lines, they can be linked to these manufacturing elements. Often a kanban (pull) system for production control is used.

6.   Unless the company has completely converted to linked-cells and completed all the steps to become an integrated manufacturing production system, they will have a mixture of manufacturing systems. Typically, most companies are production job shops with elements of flow shop where volumes permit. If they are a project shop, they will have PJS and FS making parts for the project.

7.   Trends driving companies to smaller lot production include: (1) increased variety in products, (2) shorter life cycles of products, (3) greater emphasis on quality, (4) increased variety of materials going into products, (5) increased cost of materials and energy to transform them means that the cost of inventories will keep increasing unless the size or volume of the inventories is reduced. The cost of materials (purchasing, storage, holding costs, etc.) composes 50% of the cost of manufacturing for many companies.

8.   At the turn of the century, there were fewer machining processes and fewer casting, forming, joining, and finishing processes than there are today. One of the problems with teaching manufacturing processes (as a course with restricted lab and lecture time) is that the volume of material keeps increasing every year as new processes (or variations of old processes) keep being invented to solve problems in production or deal with new materials. Many processes which can be given only a paragraph or two in this text (like wire drawing) are billion dollar industries, have entire books written about them, and have journals specifically dedicated to the subject (lasers).

9.   Group Technology is a manufacturing management philosophy which basically says that if you are making a great variety of parts, there will be certain sets or families of parts which require the same set or collection of processes. Identifying these parts allows one to devise ways to manufacture the family of parts in collections of machines (called cells), greatly reducing the through-put time, in-process storage, setup time, and labor input while greatly improving the quality. It usually takes a number of years to convert the entire job shop into a cellular shop and even then, depending on the particular business, you may want to keep some job shop operations, say for tool and die building.

10.   The manufacturing system is composed of many subsystems. Suppose that in optimizing half of the subsystems, you minimize their output and for the other half, you maximize the output. On the whole, for the system, you get an average output rather than an optimum. Part of the problem here is that the whole system is so complex, you cannot tell which parts are interacting with which other parts. If you do not know where the dependent characteristics are, then optimizing a part of the system may really be self-defeating. One of the fundamental principles of manufacturing system design is to reduce complexity of the system and the products being made in the system.

11.   A manufacturing system can be defined as a complex arrangement (i.e. a design) of physical elements characterized by measurable parameters. The physical elements of the MSD are: machine tools; material handling devices; tooling (workholding devices, cutting tools); and people (the internal customer).

12.   TPT reflects the time required for the product to move through the factory and includes all the waiting and delay time as well as processing time. See Figure 41-4. The reciprocal of the production rate, PR, is the cycle time. PR is the frequency, units/hour, for example. $1/PR = CT$ or the time between the exit of one product and the next from the process or system.

13.   Benefits include: the reduction in manufacturing costs (reduced tooling and setup, reduced inventories, improved quality, and productivity) as well as engineering designs standardization and simplification, improved materials management and purchasing, production planning and control benefits, and improved employee quality of working life.

14.   The do-nothing alternative is always an alternative that works against change. The problem is quite simple. If you are the instrument of change and the change results in failure, you have clearly failed. If you are the instrument of maintaining the status quo (the do-nothing alternative) and this course of action results in failure, you are much less likely to be branded as a failure. One of the secrets of success of the implementation of JIT/TQC is that everyone in the company, from the workers to the president, will be involved in the conversion, so everyone must succeed together or the company will fail together.

15.   An example of an A(3) machine is an automatic record changer, where playing a record is equated to the "processing " of a record. Most clocks are automatic repeat cycle devices. Examples of A(4) machines and systems are found in the toilet (feedback system control and water level in the tank) and the house thermostat. Ovens in the kitchen also have thermostats. Since there are few homes with computers tied in with the operating systems, there are few examples of homes with A(5) levels of automations and fewer still with A(6) levels. However, the day is coming when such system will be commonplace.

16.   It is difficult to model manufacturing systems because they are large, complex, probabilistic, dynamic, and nonlinear, with many dependent interactions.

17.   The route sheet is a document which describes the route or path that the parts must take in the production job shop. Each machine is indicated on the route sheet and the parts are transported from machine to machine in tote boxes or carts. The route sheet is the embodiment of the sequence of operations and processes needed to convert the part from raw material into a finished product.

18.   Line balancing refers to time balancing - making the time for each station in a transfer line as close to the same as possible. This requires the balancing of the specific operations that are done at each station so the machining time at each station is the same.

19.   The original transfer lines used hardware to actuate the tools and relays to control the hardware. This kind of automation is now called hard automation. The machines on the transfer line are placed beside a conveyor system for parts transfer from machine to machine, with no back flow. In the FMS, NC machines are placed next to a conveyor system but the parts can travel in any direction - back flow is allowed. This makes the FMS an automated job shop. In recent years, NC type machines have been placed next to transfer lines to give them more flexibility.

20.   To enhance flexibility, transfer lines are being given NC machines and PLC's.

21.   The machine must be able to complete the processing cycle initiated by the operator. The machine stops automatically and waits for the operator to return to unload the completed part and insert a new part. If the machine could not complete the cycle untended, the operator would have to stay with the machine. Thus each machine in the cell would require an operator. Some cells function this way, particularly assembly cells.

22.   In a manufacturing cell, the cycle time is controlled by the time it takes the operator to walk around the cell and perform all the necessary jobs at each machine (assuming this cell has only one worker). The machining times are less than this cycle time because they are overlapped - more than one machine is running at the same time, working on different parts. Thus the sum of the machining times can exceed the cycle time so long as no individual machining time exceeds the cycle time. This would make the machine a bottleneck and delay the CT.

23.   Flexibility is the ability to adapt to changes in the product demand (customer demand) and changes in the product design (design changes or new products).

24.   The designer can see the parts being made within the cells. All of the processes which created the part are right there together. The parts emerge from the cell as good parts or not at all. The parts pass by the machines one at a time. Quality feedback is immediate and visible. In the PJS, where the manufacturing is functionalized and spread out all over the plant floor and accomplished in lots or batches, the process capability of the collection of processes is not evident (visible).

25.   AI is short for artificial intelligence. Perhaps machine intelligence would be a better word or computer machine intelligence. I like to think of AI as the ability to infect the control software with the ability to respond to the unexpected.

26.   A feedforward device or sensor is placed ahead of the process to detect some input variable status. The input status affects the behavior of the process and thus the process is adjusted to accommodate the changing input.

27.   Robots are used in materials processing (mainly welding, painting, and machining like drilling and boring) and materials handling (load/unload castings, forgings, and plastic processing machines and machine tools for metal cutting in cells). Robots, as they become more precise with long term repeatability, are being used in electronic and mechanical assembly. Very shortly, we will be seeing dedicated robots attached to machine tools for the purpose of changing tools as well as loading and unloading functions.

28.   The typical industrial robot has a manipulator arm, a hand, a power source, and a control system. Depending upon the level of the robot, it will have feedback devices in the various joints, in the manipulator arm and hand, and perhaps sensors for tactile or visual sensory systems.

29.   Spherical, rectangular, cylindrical, and jointed-arm are common work envelopes.

30.   Positional feedback can be obtained by resolvers or stepper motors or other feedback devices in the joints of the robot arm. Proximity sensors are being developed as well to give positional feedback to the controller. Robots with visual sensors are being introduced into industry in a limited fashion. Many of these systems use vision systems for finding the position or orientation of workpieces so that the robot knows where to go to fetch the part.

31.   Instructional robots are smaller, slower, have poorer precision and poorer long term repeatability (this is not true for all small robots), and much lower load carrying capacity. They are also much safer for student use. They are usually programmed with personal computers.

32.   Digital simulation uses computer software to model the manufacturing system. The simulator can "run" the system for many days, and (perhaps) predict where the problems may crop up.

33.   The rotary transfer device not only moves the part from machine to machine, but also serves as the workholder. The robot does not serve as the workholder, only a material handler doing load/transport/unload functions. The rotary transfer machine is setup to run one part in large volume. The unmanned cell is designed to handle a family of parts in very small lots. The machines in the cell are computer NC while the machines in the transfer machine or automatic repeat cycle with no feedback -fixed or hard automation or programmable NC machines.

34.   The human worker can think, has superior vision and tactile feel, and in addition can walk. The human can also detect odors and hear funny sounds coming from the machines. Compared to humans, robots are quite handicapped but are superior in their ability to work in hazardous or dangerous or nasty environments and are very repeatable, having less cycle time variability when the cycles are long.

35.   In the football analogy, the ball is the work in progress, i.e. the material. Thus, the kicking tee would be a workholding device since it assists the kicker in the operation of the place-kicking function. It improves his distance and accuracy and replaces one worker - the holder.

36.   Tactile sensing is giving machines the sense of feel. No examples readily come to mind.

37.   Most robots in automobile body assembly lines are doing spot welding. Some are doing arc welding.

Chapter 42

PRODUCTION SYSTEMS

1.   The production system includes the manufacturing system and serves to connect and service the manufacturing system by providing it with all the other activities and services needed to allow the manufacturing system to properly function. The production system also provides information and materials movement linkages between the processes within the manufacturing system.

2.   Accurately forecasting the demand for the product is critical to the health and welfare of the company (the production system) because the manufacturing system must be able to respond to changes in the forecasted demand as well as the actual demand. No matter what kind of manufacturing system the company has, it must still forecast the demand for its products.

3.   Changes in the design early-on are relatively inexpensive to accommodate while changes in the design after the product is being built can be very expensive. The more knowledgeable the designer is of the capability of his manufacturing processes, the more likely he is to design things that can be readily built by the system. This is called designing for producibility. Designing something to be made in a totally automated manufacturing system requires a totally different design from the same item built in a manual assembly system. Jobs done easily by people may be impossible for even the most sophisticated automatic equipment and vice versa.

4.   For example, the work boot of a foundry worker has steel toes and a strong arch to protect his toes from heavy objects being dropped on them and provide good support on the hard concrete floors of the foundry. One doesn't wear sandals in the foundry as these are designed for entirely different purposes.

5.   The overhead costs include all those costs necessary to run the factory but which are not tied directly to the product. The cost of the foreman or the forklift truck drivers, power, light, heat, indirect materials, and so forth are all totaled into the overhead cost.

6.   A significant part of the overhead cost will be in the cost of work-in-process (handling and storing of inventory), raw materials storage, and floor space needed to store such inventories. Not insignificant will be the cost of performing rework of parts not made right the first time (the cost of the hidden factory). Better planning and scheduling can reduce these costs and thereby reduce the level of the overhead.

7.   The design determines which manufacturing alternatives will be available to make the part. The design, along with the needed volume and the material selected for the part all influence the choice of the manufacturing processes. For example, standardiz-

ing the design of the thread type and hole size greatly simplifies the design of manufacturing cells. Suppose the design calls for 16 RMS finish. The manufacturing system will probably need a grinding operation to meet this design specification. The manufacturing cell design would have to include grinding capability. Design also influences the production system in many ways. Designing things that customers want to buy, that can be readily inspected for quality, that are reliable, and that are safe for the customer to use are all design aspects that impact the production system. Because design occurs before all the other functions described in the manufacturing and production systems, it obviously is the driving force.

8.   This statement means that a large expensive piece of software adds a large fixed cost to the total cost of making something in the same way a large expensive piece of hardware adds a large fixed cost to the total. Both costs will require a large volume of parts to be made to cover the cost. The problem is that while hardware (equipment) can be depreciated and some of its cost recaptured through tax savings, software costs are not depreciated and are generally hidden in the overhead costs of the company.

9.   Many processes are specifically designed to deal with specific materials. In the USA, we have designed our technologies so that welders are not trained or allowed to be metal formers and machinists are not allowed to be welders. That is, our systems and processes are single functional and thus inflexible. A technology like wire drawing is not suitable for cast iron, but is for steel. The drawing process introduces into the material a specific deformation history, which is carried forward into the follow-on processes. An extruded bar will have a different history and will thus behave differently in follow-on processes than a drawn bar.

10.   He has eliminated permanent mold casting and die casting, and some of the sand-based techniques.

11.   If the chef has no charcoal broiling pit, he can not offer steaks and ribs cooked by this means. To do french fries requires him to have a deep-fat fryer. What he can put on the menu depends upon the cooking facilities he has available. The size and capacity of the restaurant will be limited by the capacity of the kitchen.

12.   Many of the case studies are good examples of situations where improper design details lead to part failure. It is more common that failures in service are a combination of design error, material problems, and fabrication problems all tied together and often compounded by improper usage by the customer. One of the grave dangers of computer-aided design is that design will be done by the computer in the hands of nondesigners, leading to poor designs which are rapidly and efficiently done, but improper none the less. Obviously the same can be said for CAM. The best manufacturing engineers and the best designers are

the ones who should be doing CAD and CAM. Perhaps we should
write these acronyms D,ca and M,ca so that we remember that the
computer is just an enabling technology.

13.   The flange would be a flat disk and the shaft would not
have any steps in it.

14.   Both designs provide a step on the shaft to seat the
flange. The improved design, however, eliminates some machining
in the flange but calls for registration against the outside
surface of the flange, which may not be a machined finished
surface. Thus, the accuracy and precision of the mated parts will
be dictated by the overall thickness of the flange and not the
step inside the hole in the flange. Also, if these parts are to
be mated by a shrink fit, the thinned down wall of the shaft and
the sharp corner radius at the thinned down wall may cause a
failure under shrink fit loads.

15.   Yes, an NC lathe could make this part as could many other
turning machines, including a turret lathe and a single spindle
automatic. Whether the part would be made on a more efficient
machine would depend upon other factors not given in this
problem, including the volume. (Note that this assumes that the
center drilling of the ends is being done just so the part can be
held between centers.) For 25 parts, an NC lathe would probably
be justified in terms of cost.

16.   Insufficient volume is the only thing that prevents one from
doing this part on an automatic lathe. Note that both ends of
the part are center drilled. If this operation is needed for
something other than holding the part between centers, then the
automatic lathe might not be the right choice.

17.   The blades are pierced and blanked out of strips of steel
and passed (as a strip) through a series of grinders and hones
(and even leather strops) which put the cutting edge on the
blades. The blades are then separated into individual blades by
a shearing operation, stacked on holders, and placed in
sputtering or coating processes (in large bulk qualities) for
chromium edging and plastic coating (teflon is used to reduce the
cutting friction). The blades cost about 2.5 cents each.

18.   The left end which mates with the die and the right end
which mates with the hole in the upper portion of the die set are
the critical elements. The overall length is not critical
because the punch passes through a hole in the die to blank and
pierce and a couple of thousandths one way or the other are not
critical.

19.   For a lot size of 250, an NC machine would probably still
have been the best choice. In the main, turret lathes have been
displaced on the factory floor by NC machines. For 25,000 parts,
a single spindle automatic machine or screw machine would be a
good choice.

20.   The student should provide location dimensions for the hole in two directions, x or the horizontal, and y or vertical, and a radius dimension for the curved surface.

21.   Yes, these dimensions are within the capability of extrusion and cutoff assuming the cutoff is being done with one of the new horizontal band saws. If the hole is extruded, then there will be some burrs in the hole. If the hole is not extruded, then it will have to be drilled and perhaps reamed.

22.   The question will be answered by students with examples of products whose design has been changed to enable them to be automatically assembled. The stores are full of them.

23.   The initial shape could be produced by casting or by welding two plates A and B together with a large fillet C. The holes would need to be drilled and the slot milled. The sequence of operations cannot be specified since no dimensions are given.

24.   An operations sheet gives more specifics with regard to the processes needed to make the part while the route sheet, a production control device for the job shop, provides information about where the part is to go next for more operations. In the visual factory, operations sheets are posted right at the machine for all to see.

25.   The question requires the student to go to a factory (or even a fast food restaurant) where operations sheets are used, get a copy of one, and bring it back to the class and explain it. Who designed the operations sheet itself? Who fills it in (determines the details of the operation)?

26.   Ergonomics studies all aspects of the interaction of humans with machines and includes fatigue related injuries, biomechanics of work, safety, and other human factors.

27.   Motion and time study, methods analysis and video taping the setup are the most commonly employed techniques.

28.   The AQL stands for Acceptable Quality Limit and refers to the % defective that is acceptable in a sampling type of inspection scheme. The modern approach to Total Quality Control says that no % defective is acceptable.

29.   The dispatcher initiates or dispatches a production order. The expediter finds jobs which have become delayed for one reason or another and gets them moving again.

30.   Production control refers to controlling the movement of the materials to the right machines at the right times. Production control deals with when to make the products (scheduling) on which machines in what quantities. Inventory control deals with having the right amounts of materials in the system available at the right places at the right times.

31.   If I am manufacturing bikes, the demand for tires is
dependent on the number of bikes I make - two tires per bike.
The number of bikes is independent and depends on customer
demand.

32.   The mrp generates orders for the shop, which generates the
orders for purchased parts (from the vendors) and the orders for
subassembly and component manufacturing. The MPS uses the
information in the BOM as one of its inputs. The BOM lists all
the parts that are in the product. The MPS uses the information
regarding the capacity of the systems compared to the orders for
the components and products to generate a master schedule.

33.   See Figure 42-19

34.   Finite element analysis is a method to determine the stress
distributions with a part under various loading conditions, the
strains in the part, and the temperature distributions in the
components.

35.   The variant method uses variations of existing process
plans to get the plan for the new part. The generative method
generates a process plan from scratch. Hopefully, both of these
tasks are done by the human with the aid of a computer.

36.   MAP stands for Manufacturing Automation Protocol.

37.   LAN stands for local area network.

CASE STUDY - CHAPTER 42

The Underground Steam Line

1). and 2). The holes (cuts or slits) in the outer casing
occurred where the edges of the U-shaped supporting legs
contacted the casing. Essentially the entire weight of the outer
casing rested on the upper vertical leg <u>during transportation</u>.
During operation, the weight of the inner line and the steam will
rest on the two lower legs. The designer of this line was not
sufficiently familiar with cold forming operations to realize
that when flat bar stock is bent to form the U-shaped support
legs, a sharp corner will be produced on each outer corner of the
bent pieces due to the stretching-contraction of the bars about
the center line. This will create a line contact where the legs
contact the outer casing, and the weight of the casing will be
great enough to cause the load per unit area to exceed the
strength of the casing wall in these regions. This was a design
error. (The designer never went out to look at the line being
fabricated or he might have caught this condition at the plant.)

3). Two design modifications are suggested. Grind off the sharp
corners of the legs and shape the contour of the leg to conform
to the interior curvature of the outer casing. This will reduce
the unit loads to an acceptable level. The second design

modification might be to eliminate the legs altogether and let the steam line simply lie on the bottom of the return line with a layer of appropriate insulation material placed between the two. This eliminates the legs and their fabrication and installation altogether.

4). The line must be totally disassembled and these modifications made even though no damage was visible or detectable. An entire new outer casing should be obtained since it appears that the damage was caused during shipping, not installation. Once operational, however, similar damage to the outer casing will be produced by the bottom legs, unless modifications are made.

Chapter 43

LEAN PRODUCTION: JIT MANUFACTURING SYSTEMS

1.   When the machines are placed together into interim manufacturing cells, they must be able to switch quickly from one part to another. Therefore, setups must be very rapid or else too much time will be lost doing setups.

2.   The OHNO system builds in small lots throughout the system. The FORD system initially did the same thing but as the product variety grew, so did the lot sizes or the batch size until only the final assembly line built in small lots. The rest of the system built product in large lots.

3.   The U-shape design of the manufacturing and assembly cells places the workers in close contact with each other so that they can help each other. One of the ideas that fosters high-touch is putting the workers close enough that they can hand each other parts directly and assist each other when the need arises. The workers handle the parts between every step in the process, so high touch is very evident in the manned manufacturing and assembly cells.

4.   S.M.E.D. is an acronym for the single minute exchange of dies. This term was coined by Shigeo Shingo, an I.E. consultant who worked with Toyota to reduce setup times. Single minute really means single digit, i.e., less than 9:59.

5.   The functions of the production system are integrated into the manufacturing system. The functions of quality control, inventory control, production control and machine tool maintenance are integrated into the manufacturing system, meaning that they are done as part of the routine daily operation of the lean production system.

6.   The structure of the linked-cell manufacturing system defines all the paths that all the parts, components, subassemblies, and final assemblies take during manufacturing. All the elements are linked together by Kanban or directly. The final assembly line(s) pull all the parts and subassemblies together as needed, when needed. In other words, the timing of the manufacturing system is an integral part of the system.

7.   The worker that can operate more than one kind of machine is a multiprocess worker. He can run a lathe, a milling machine, and a jig borer. If the worker is multifunctional, he can perform other functions besides machine tool operation. He can perform quality control duties, reduce setup times, maintain the machines in a routine way, perform routine housekeeping of the area of the plant he works in, and work to continuously improve the processes in the cell.

8.   Integrating quality control into the cells begins with getting the worker to take responsibility for the quality of the product and getting everyone to begin to inspect products to prevent defects from occurring rather than to simply find the defects. The people that make the products are responsible for the control of the quality of the products. The adversarial relationship between inspectors and production workers is eliminated, as everyone is an inspector.

9.   The manufacturing systems run at a normal, regular pace rather than a big rush at the end of the month or the end of the week. The systems have reserve capacity that can be used to handle emergencies and increased demand. The processes run without undue pressure at speeds which permit long tool lives, no machine tool breakdowns, and unpressured workers.

10.   Workers from other areas come and pick up carts which are full and ready for use. These workers use a WLK to do this and leave an empty cart in place of the full cart. They detach a POK from the full cart and put it in a POK collection box beside the work area. The order of the cards in the collection box defines the order that things will be made in the cell. "What to make" has been previously defined by the family of parts that this cell makes every day, day after day, routinely. Therefore, the pull system of Kanban tells what and when to make certain products.

11.   The final assembly line is designed to run in a mixed-model fashion rather than running batches of products in set sequences. Mixed-model final and sub assembly means that the line makes all the products one at a time in any order. Suppose you are making products A and B on the line. Instead of making 20 of A, then 20 of B, you make them ABABABABAB etc. Since the final assembly pulls parts from the rest of the factory, this scheme levels out the rate that parts are needed. Leveling is also called smoothing the production.

12.   Suppose that between two manufacturing points there are 10 carts with 20 parts in each cart. The inventory between each point is known - 200 parts. The foreman in the area takes a set of Kanban cards out of action. This puts a cart of parts on hold. The cart and its inventory cannot be used. If a problem appears, the inventory is restored and the problem worked on until it is solved. The foreman repeats the process, time after time, until no problem appears. In this way, the inventory is controlled between any two points in the system and is continuously reduced. Note also that the inventory level becomes a control variable for uncovering problems.

13.   The inventory is used to cover up problems of poor quality or machine breakdowns. The longer the setup time, the larger the lot size needed to cover the long setup time and the larger the inventory.

14. The length of the setup times, the number of defective parts, the frequency of machine tool breakdowns, the travel distance between cell and use point, and the frequency of occurrence of other problems determines the minimum level of inventory in the link. If all these kinds of problems have been eliminated, the inventory level can be very low.

15. Once you have designed and implemented a factory of linked-cells (steps 1 through 7), you can expect that each of your vendors would be prepared to do likewise. Therefore, each vendor become a manufacturing cell but in remote location. The vendor tries to ship daily to the factory to keep the inventories at a minimum. This may not always be possible or economically advisable.

16. <u>Autonomation</u> means to prevent bad material from passing on to the next operation automatically or to automatically stop a process when it has made the necessary quantity in the necessary time. The object is to not make any defects, or to not pass defective material on to the next process or to not make unnecessary parts (do not build to inventory, just to needs). It is a key concept to producing zero defectives or 100 percent perfect parts. The closer you are to producing 100 percent perfect parts, the closer you can get to a lot size of one, which is the ideal lot size for the JIT system.

17. JIT or lean production is a strategy that restructures the shop into linked-cells and integrates the critical production functions into the manufacturing system, then computerizes. CIM strategy does not try to reorganize or restructure the job shop but to computerize the existing structure. Even the most ardent CIM proponents are beginning to admit that first you need to become lean, then you can computerize the system.

18. The CT = 1/PR where PR is the production rate.
   PR = Daily Demand/Hours in Day
   Daily Demand = Monthly Demand/Number of working days in
      month.
   This CT applies to the entire facility, including the
      manned manufacturing cells.

19. The more workers in the cells, the shorter the cycle time as long as the CT is greater than or equal to any automated individual machining time or assembly processing time.

20. Concurrent design is accomplished with a design team which looks at all aspects of the manufacturing at the same time the product is being designed.

21. PFA is a method of Group Technology called Production Flow Analysis. It uses the information from the route sheets to find families of parts and groups of machines to make the parts. The hard work is to design the cell and the tooling to make the parts.

22.   Your ZIP code tells where you live (in a general sense) as does your telephone area code.

23.   In the manufacturing cell, the worker moves from station to station or machine to machine. He is not tied to a specific machine but rather operates many machines.

24.   A robotic cell is not as flexible as a manned cell because you have removed the most flexible element - the man. The robot has one "hand", poor senses compared to humans, can't walk very well, and isn't very smart.

25.   No, the robotic cell is not the same thing as an FMS. The cell functions like an assembly line with unidirectional flow -no back flow. The FMS functions like the job shop, taking parts in random order and direction of movement.

26.   In the SMED system, internal setup items are those elements which require the process to be stopped in order to do them. External setup elements can be done while the machine is running.

27.   The intermediate jig interfaces with the machine tool, and all the intermediate jigs look identical to the machine tool. The workholding devices are specific to the parts and are mounted on the opposite side of the intermediate jig. A cassette for your tape player is an intermediate jig. All the tapes are different but to the tape player, all tapes look the same. An intermediate jig is standardized in terms of its location.

28.   The SMED methodology is taught to all the production workers, foremen and supervisors. Everyone on the plant floor does setup reduction. If you wait for the setup reduction team to come around to all the processes, the job will never get done.

29.   The production workers are taught how to control the quality of the parts being made in their processes by learning how to control their own processes. They are taught how to locate assignable causes of problems. Defective items are not permitted to leave the cell.

30.   Quality circles are a form of teaming with the specific objective being to improve quality and solve problems related to the processing of the products. QC's are considered to be an on-line technique.

31.   World Class Manufacturing, JIT/TQC or JIT manufacturing, the Toyota Production System, the OHNO System, MAN system (Harley Davidson), and other names specific to companies.

32.   See response to question 12 in this chapter. The WIP is measured by the number of carts between two production points times the cart capacity. WIP is reduced by steadily removing carts or reducing the cart capacity which in effect reduces the

inventory level between the two points and exposes problems. When the problem is solved or removed, the inventory is further reduced.

33.   Taiichi Ohno was the Vice President for manufacturing at Toyota and is generally considered to be the father of the famous Toyota Production System.

34.   JIT purchasing has many differences from other purchasing systems but fundamentally the idea is to get one very good vendors, to work closely with this vendor so that he will make superior goods for you at the lowest possible cost and that he will deliver these items to you when you need them in the quantities you desire and you will not have to inspect them when you get them. That is, the quality will be as good as it possibly can be -- perfect if possible.

35.   The constraints include the fact that systems are hard to change. It's easier to cost-justify new equipment than it is to justify the cost of moving the old equipment to a new layout. The decisions makers fear the unknown. The new system is the unknown. The agents of change have the most to lose while the agents of the status quo have the least to lose. The middle managers have the most to lose as many of the functional areas they represent will be restructured in lean production.

PROBLEMS FOR CHAPTER 43

1.   a)   1 min., 50 Sec. = 110 sec./2 = 55 sec.
      b)   118 sec.
      c)   Yes. 480 min. per day/.9167 min per part = 523.6 parts
            per day
      d)   The long machining time for the finishing process means
            you have to duplicate it in the cell and visit each
            machine every other time you pass through the cell.
            Otherwise, the finishing process would be a bottleneck
            process and delay the CT for the cell.
      e)   55 - 42 sec./55 = 13/55 = 23.6% of the CT is spent
            walking.
      f)   The walking time can be reduced by designing machine
            tools that have very narrow footprints (widths) so the
            operator can move quickly from machine to machine.
            Manufacturing cells that use machine tools designed for
            stand-alone (job shop) manufacturing have large foot-
            prints because the worker is assumed to be stationary.
            Simulation can be used to find layouts that reduce
            operator walking distance. We have used a technique
            called "paper dolls" where we make cardboard cutouts of
            the machines and put them down on the floor and have
            the operators determine the "best" layout.

CASE STUDY - CHAPTER 43

The Case of the Snowmobile Accident

This case was fictionalized from an actual case in which one of the authors of the text was involved as an expert witness. In real cases of this sort, there will always be conflicting evidence and many explanations presented for the same set of evidence. Often, the truth may never be presented to the jury and the actual cause of the accident may never be known. Such was the situation here.

The tierod sleeve did actually fail prior to the accident. The cause: the tierod sleeve broke under impact when the right ski of the snowmobile hit a deep rut. At twenty below zero, low carbon steel has low impact strength and behaves as a brittle material. The original designer erred in his choice of materials for this application. The failure was not due to the reasons the lawyer stated, although in combination, such was certainly possible. Nor did the tierod break when the snowmobile hit the tree, although this was also possible.

In cases like these, there are frequently no winners, except the lawyers.

APPENDIX CASE STUDIES

APPENDIX A - CASE STUDY
Heat-Treated Axle Shafts

The substitution of the plain carbon steels for the previously used alloy steels was made possible by the use of a scanning induction-heating coil followed by a severe water-spray quench assembly. Only a small section of the bar is hot (and therefore weak) at any given time, and it is supported by cold, rigid material on either side. Moreover, if the coil scans a vertically oriented axle, gravity maintains alignment and retards warping. A symmetrical quench unit surrounds the bar to provide a uniform, rapid cooling. These bars will not through-harden due to inadequate hardenability. The final structure will be a hardened martensitic surface with a softer, more fracture-resistant core - a very desirable combination for an axle. The martensitic surface can then be tempered by another scanning induction-heating coil with a power input designed to produce the lower tempering temperature.

To achieve a surface hardness of Rockwell C 60, the steel would have to have a carbon content of 0.5% or greater. Thus, the 1038 or 1040 steels must be upgraded to a higher carbon 1050 grade (using the lowest carbon content that will provide the desired properties). The hardened bearing region can then be obtained by a second induction heating of just the surface of the desired region (higher frequencies heat to shallower depth) by a stationary coil, followed by a severe water quench and selected induction temper. By heating only the outer skin, rigidity is maintained to prevent distortion. In addition, the quenching then sets up a favorable residual stress pattern with the bearing surface in residual compression.

Alternative methods of hardening the surface, such a carburization, would be unattractive since they would modify not only the desired bearing region, but also the entire remaining section of the shaft. Selective carburization would require some barrier be applied to the remaining regions, such as a copper plating, and this imparts additional expense.

APPENDIX D - CASE STUDY
Handle and Body of a Large Rachet Wrench

1). Because of the size and shape of the part (i.e. no constant cross section or rotational axis of symmetry), plus the presence of the through-hole, it would appear that this part would be a candidate for processes such as impression-die forging (either drop hammer or press), or some form of casting. Depending upon the material selected, casting alternatives might include die casting, permanent mold, lost foam, shell mold, or even investment. Conventional green sand casting would probably produce an unacceptably rough surface finish that would require secondary surface processing. Depending upon the process, it may be necessary to machine or finish machine the through-hole. Since this hole has a complex cross-section and angular corners, a process such as broaching may be preferred over techniques such as milling.

2). Because of the requirements of strength, corrosion resistance, and light weight, the part appears to be a natural for some form of aluminum alloy. Titanium would appear to be overly expensive. Copper alloys are heavier than steel, and steel would need some form of plating or coating to provide the necessary corrosion resistance. Stainless steel would meet the requirements, but is more difficult to fabricate than aluminum, is more expensive, and is approximately three times heavier (NOTE: the wrench is nearly 2-foot long!). The die-castable 27% aluminum zinc alloy (ZA-27) can also meet the desired properties.

3). If forging is preferred, the selection will probably involve some form of age-hardenable wrought aluminum alloy, such as a 2xxx or 7xxx series alloy. If casting is the selected process, an age-hardenable casting alloy would be preferred, such as alloy 201.

4). The selection of "best" solution is really one of judgment. From a metallurgical perspective, forging might be preferred over casting because the combination of rolled barstock as the starting material with the added flow of forging would orient any flaws or inclusions in an axial direction where they would not tend to act as crack initiators or crack propagators. With this solution, however, the hole could only be made as indentations from both sides with a center web that would have to be punched out and finish machined.

5). For aluminum alloys to possess the desired strength, an age hardening heat treatment would probably be required. An anodizing (or possibly color anodizing) treatment may be used to enhance surface properties.

6). A safety tool would generally have to be made from a non-pyrophoric material (one for which small chips or slivers would not "burn" or oxidize in air). Aluminum is a pyrophoric material, but aluminum oxide is non-pyrophoric. Therefore, if one could assure that the surface has a sufficiently thick oxide layer, it may be possible to use the aluminum alloys selected above. There are standard tests that a safety tool must pass, and it is possible for our aluminum product to be accepted. If a non-pyrophoric material were required, it may be necessary to forego the requirement of light weight and use a copper-based alloy. Since the strength requirement is not exceptionally high, there does not appear to be a need to use copper-beryllium alloys (a popular material for high strength safety tools), for which toxicity concerns are significant.

APPENDIX C - CASE STUDY
Diesel Engine Fuel Metering Lever

NOTE: This problem is particularly attractive because of the large variety of material-process combinations that can meet the required geometric, physical, and mechanical properties.

1). As is usually the case, the part could be fully machined from a larger piece of metal, such as a rectangular flat. This, however, is usually an inefficient use of material and time and labor considerations may be restrictive, especially for a production run of 20,000 pieces.

The small size of the part, smooth surface finish, and presence of a through hole, make the part attractive for one of several casting processes, including investment, die, and permanent mold. In addition, the part might even be attractive for centrifuging.

Noting that many of the surfaces are flat and parallel (such that if the part were viewed along the hole axis, the cross section would be uniform), one might want to consider extrusion of a shaped section, rolling of a shaped bar, or cold drawing, plus machining to remove the unwanted portions of metal. This would significantly reduce the amount of machining from the first alternative in this section.

Finally, the small size and prescribed wall thicknesses render the part a candidate for powder metallurgy, or P/M injection molding, as a means of production.

2). These properties are not very restrictive, and can be met by a number of metals and alloys, both ferrous and nonferrous. These include most steels, ferrous P/M alloys, copper-base alloys, some heat-treatable aluminum alloys, zinc-aluminum die casting alloys, and a variety of others.

3). The processes listed in part one include both wrought forming and casting, as well as powder metallurgy. This section is designed to get the student to focus on process-material limitations. Almost any metal can be machined, but if 100% machining were to be employed, a free-machining metal or alloy should be seriously considered. Of the casting processes, investment would be the slowest and most costly. This would probably only be considered if ferrous materials were required. Since alternative metal systems can provide the desired properties, die casting of either a copper-base or zinc-aluminum alloy would be an attractive alternative. Ferrous materials cannot be die cast and the higher-melting-point copper-base alloys have a limited die life, so the zinc-aluminum alloys might be preferred here. Extrusion would require a ductile, wrought alloy, such as an age-hardenable aluminum. Cold drawn bars of low carbon steel would also meet the requirements and copper-base alloys might be considered here. If powder metallurgy were pursued, a ferrous powder would likely be required, but the low hardness and ductility requirements provide ample room for such a solution. The complexity of shape might lead to a preference for P/M injection molding over the conventional press-and-sinter powder metallurgy approach.

4). The conclusion as to which solution is best is indeed a question of "multiple shades of gray". Each of the above possibilities has merits, and the "best" solution may well be based on the experience, available equipment and expertise, and current economics of the various processes and materials.

In each case, the form of the starting material would be different. Full machining would begin with mill-length bars of standard configuration. Casting would begin with melt-quality ingots. If using extrusions or complex cross-section bars, the primary forming operation would likely be contracted out to a specialist firm and the product could be purchased with a specified degree of cold work amenable to both finish machining and final properties. Powder metallurgy would begin with a specified blend of powder and lubricant.

The necessity for heat treatment again depends upon both the material and method of manufacture. Some of the above systems would require age hardening to attain the desired final properties. Ferrous P/M would require a quench and temper. Other alternatives could meet the goals with cold work.

APPENDIX D - CASE STUDY
Bevel Gear for a Riding Lawn Mower

NOTE: While this problem is generally restricted to the stronger ferrous materials, the size and shape offers multiple means of fabrication.

1). Based on the size and shape of the part, possible means of production would include forging from barstock, casting, conventional powder metallurgy (press and sinter), and possibly even machining from rolled plate. Because of the demands of a gear, fatigue and fracture resistance are additional demands. For these properties, it is desirable that flaws or defects not intersect perpendicular to critical surfaces or locations, such as the base of the individual gear teeth. From this perspective, the worst process would be machining from rolled plate, since all flaws would be oriented along the rolling direction which is within the plane of the gear. By forging rolled barstock, defects would be oriented in an axial or radial alignment! Because of the maximum thickness of only 1/2-inch and the fairly uniform thickness, powder metallurgy would be an attractive option, provided the density could be made sufficiently high to impart adequate fracture resistance.

2) and 3). The requirements of strength and hardness point to some form of ferrous material. If fabrication is by forging, then a plain carbon or low-alloy steel would be preferred. If casting is selected, steel offers more difficulty than acceptable types of cast iron, such as ductile or malleable grades. Austempered ductile may be particularly attractive. Powder metallurgy offers a variety of unique alloys (since blending is accomplished in the solid state). For these requirements, one of the standard iron-nickel powders may be attractive.

4). Again, the "best" solution is often a matter of preference. This part has been commercially produced by forging, casting and powder metallurgy, and all products have performed in an acceptable manner. The desired "condition" of the material may be of particular significance to the metallurgical engineer. The specified properties can be met by a full range of conditions, from air-cooled through quench-and-temper. If sufficient consistency could be achieved in the finishing temperature, forgings might be used in the as-forged condition (simply air-cooled from forging). This would eliminate the handling, expense and time required for a subsequent heat treatment. Concerns should relate to both the desired structure and the consistency of the structure and properties from part to part.

5). The production of a gear blank and the subsequent machining of the teeth has been a standard means of producing gears for a number of years. However, the machining operation adds additional steps to the manufacture, along with the associated expense of handling, fixturing, cutting tools, material waste (chips), lubricants, and others. A more expensive initial operation that could eliminate the secondary machining might be a less costly means of producing the final part. Because of problems associated with scaling, warping, and phase transformations, high elevated temperatures are generally not attractive when trying to form ferrous materials in a net-shape process. However, the associated softening may make warm forging an attractive alternative. The presence of some moderate degree of strain hardening may remove the need for subsequent heat treatment, provided the initial process can be adequately controlled to provide the desired consistency. A surface treatment, such as shop peening, can be used to clean the part, produce an attractive texture, and impart additional fatigue resistance.

APPENDIX E - CASE STUDY
Rocker Point for an Electronic Scale

NOTE: Because of the low mechanical properties and the nature of the size and geometry, this part offers a wide range of manufacturing options.

1). Because of the size, shape, and presence of two sets of perpendicular holes or slots, this part can me made in a variety of ways. Die casting with retractable cores would appear to particularly attractive. Alternate means of casting would include permanent mold, low-pressure permanent mold, and investment. Powder metal injection molding is another possibility, but the large amount of shrinkage observed during burn-out and sintering may make it difficult to achieve the necessary precision. Because of the parallel surfaces, both horizontal and vertical, the part could be machined from a rolled or extruded bar that would have the desired cross-section in one of the two directions. The specific means would probably be based on minimizing the amount of material to be removed by machining. Since the mechanical properties are sufficiently low, some form of polymer molding process may also be a possibility.

2). Because of the relatively low mechanical requirements, coupled with the absence of a hardness or impact specification, there exists a wide variety of material possibilities. Many of the engineering metals, including zinc, aluminum, copper and iron-based metals would be possibilities. A number of ceramic materials would be adequate, as would some of the engineering polymers.

3).-5). An extremely wide variety of material and process combinations are available, such that it would be difficult to cover the breadth of even reasonable possibilities. The instructor should note that problems such as this can often form the basis of in-class discussions relating to the relative advantages and limitations of various alternatives, with students defending their own solution and criticizing those of others.

APPENDIX F - CASE STUDY
Automobile Water Pump Impeller

1). This is another part that can be produced in a variety of ways. As designed, the part is a two-level part with flat surfaces. This, coupled with the relatively small surface area and small thicknesses, would make the part attractive for manufacture by conventional press-and-sinter powder metallurgy using a double-action press. Alternative means of manufacture would probably involve some form of casting, such as die casting, permanent mold, shell, or investment. It would be difficult for forming processes to produce the existing design because of the lack of draft or taper on the impeller blades. With design modifications, impression-die forging might be a possibility.

2). The relatively low mechanical properties, the low ductility, and the absence of a hardness or wear requirement make this part a candidate for a variety of materials. Because of the presence of coolant (an electrolyte material) and additional materials in the shaft and housing of the pump, material selection might be based as much on galvanic corrosion as on mechanical performance. Possible materials include aluminum, cast iron, copper alloys, stainless steel, and others.

3). - 5). Again, the spectrum of possibilities is great. If conductive material (i.e. a metal) is specified, consideration should be given to galvanic compatibility with what will likely be a steel shaft and the material to be used in the housing. Heat treatments would not likely be to produce enhanced strength, but may be specified to effect a stress-relief. Surface treatments might be such as anodizing, if aluminum were specified.

6). Since this part will be constantly exposed to water-based solutions over a range of temperatures, the response of polymers to water immersion would be a major consideration. Many polymers absorb water and exhibit dimensional swelling. By proper selection of resin, and the use of appropriate fillers and/or reinforcements, it would appear that a polymeric solution to the above requirements would indeed be feasible. Fabrication would be by one of the polymeric molding techniques. By selecting a nonconductive polymer, galvanic concerns would be removed, and the major concerns would now relate to mechanical durability - resistance to swelling, cracking and erosion.

APPENDIX G - CASE STUDY
Flywheel for a High-speed Computer Printer

NOTE: This part is somewhat unique in the relative absence of mechanical requirements (strength, ductility, fracture resistance, etc.) and the need to concentrate high mass in a small part (i.e. the desire to use a heavy material).

1). The above requirements, coupled with the need for high dimensional precision tend to restrict the possibilities. Since all axial surfaces are parallel, and the presence of gear teeth and a non-circular section add complexity in this plane, powder metallurgy seems extremely attractive. The thickness of the part is rather high for powder metallurgy, but the mechanical properties are sufficiently low that the absence of high-density pressing should not be a major limitation. The processing of ferrous materials has become routine for powder metallurgy, and heavier copper-based alloys could be used if even greater mass is desired within a given shape. Machining from bar stock would be another alternative, especially in view of the specified precision. Casting processes could be considered, but those that are compatible with ferrous materials would likely require secondary processing to attain the desired dimensional precision.

2). Since the mechanical properties are largely unspecified, powder metallurgy is free to select a material based on minimization of expense and ease of fabrication. An unalloyed iron powder, possibly an iron-carbon to utilize the benefits of the graphite as a lubricant, would seem attractive. Another alternative might be the iron-copper powders, since copper additions enhance P/M fabrication and the copper will actually add additional mass, possibly off-setting the presence of voids within the P/M product. Fabrication by machining would probably utilize some form of free-machining steel bar. Casting processes would best utilize one of the more fracture-resistant cast irons, since the part is a spinning flywheel, and the brittleness of the cheapest gray cast iron may be a detriment.

3). The "best" alternative here appears to be powder metallurgy because of the suitability of the size and shape, the low or absent mechanical properties, the desirability of ferrous material, and the minimization of scrap and labor (compared to machining). Fabrication would be by the conventional press-and-sinter method.

4). Consideration might be given to the need for enhanced corrosion or wear resistance. The part contains numerous small gear-type teeth -- might they experience wear. If ferrous material is used, is corrosion a possibility. If either of these become a concern, surface modification, such as the popular steam treatment of ferrous powder metallurgy products might be considered. Such a treatment would enhance surface properties without significantly altering the surface dimensions or finish.

APPENDIX H - CASE STUDY
The Broken Wire Cable

The cable manufacturer is correct. The cup-and-cone fractures and evidence of necking indicate that the wires had been overloaded and had exhibited normal, but limited, ductility. This is precisely the type of behavior that would be character- istic of a wire that had been cold-drawn through dies and not subsequently annealed.

A wire cable is not likely to be defective when new because such a cable is composed of many individual wires, that probably originated from multiple coils of rod and, quite likely, from more than one heat of steel. Thus, there is little possibility that more than one wire would be defective at any given location. Moreover, the fact that each strand of wire has withstood the tensile stresses of the wire drawing operation somewhat assures the quality of each strand.

It is possible that the cable being used is not the proper cable for the job and is indeed being used above its rated capacity (or the equipment was being abused by applying excessive loads). In addition, it is possible to damage such cables during use by such acts as: using pulleys or sheaves with too small of a bend radius (thereby putting some strands in tension and others in compression as the cable is bent), using pulleys or sheaves with grooves of improper size and shape, permitting the cable to sit static under load with a certain region of the cable experiencing a small bend radius, providing inadequate lubrication to the cable (thereby suppressing the relative movement of strands as the cable is flexed), and others. Most of these mechanisms, however, would produce additional marks or evidence on the cable, such as wear marks, reduced cross section, nonsymmetric deformation, etc. In the absence of this additional evidence, one would conclude that the failure mode is most likely a static overload applied to the cable through the crane.

APPENDIX I - CASE STUDY
The Short-Lived Gear

1).   The heat treatment sequence appears to concentrate on surface hardness and fails to consider the need for altering the interior properties. The subsurface regions of the gear retain the soft, full-annealed structure that was imparted for easy machining. The failure mode was most likely a deformation of the weak, interior metal. This could be confirmed in several ways. By sectioning through the teeth, specimens could be prepared for metallography and microhardness. Metallography should reveal the coarse pearlitic structure imparted by the full anneal, and the failed gear should show signs of plastic deformation. A microhardness scan from the surface inward should reveal the drastic drop in hardness (or strength) as one moves inward from the surface. Bulk hardness testing would also confirm the presence of a soft interior.

2).   Proper manufacturing would require the insertion of a hardening treatment for the entire gear after machining, followed by a tempering operation, and then the surface hardening treatment. This may still fail to produce an acceptable product, because the original material, 1080 steel, was apparently selected to provide the high carbon necessary to produce the desired level of surface hardness. The 1080 alloy may lack sufficient hardenability to form a tempered martensite structure in the interior, and the high carbon steel may provide insufficient toughness or fracture resistance in the core.

A more attractive alternative would be to select a low-carbon alloy steel, with the alloy content being selected to provide adequate hardenability for the specific dimensions of the gear. The material would then be annealed and machined, as done previously. The gear-shaped product would then be carburized, to raise the level of carbon in the surface of the gear. The product would then be subjected to an austenitize-quench-and-temper heat treatment to impart the desired final properties. NOTE: If higher hardness is desired in the teeth, a flame or induction hardening could be applied to the carburized surface.

APPENDIX J - CASE STUDY
The Broken Marine Engine Bearings

1). The facts point to an inadequate quenching treatment. For this material (high-carbon alloy steel), completion of the martensite transformation will not occur until the quench attains temperatures below room temperature. If a room temperature quench is used, followed by the normal tempering, a significant amount of retained austenite can be present in the bearing. Subsequent cooling to below the quench temperature can cause the retained austenite to transform to untempered martensite, causing the material to become brittle and bringing about a concurrent expansion in volume. This expansion may be sufficient to cause the engine to seize. An alternative mode of failure would be fracture of the brittle, untempered martensite.

Confirmation of these suppositions could be obtained through metallography, X-ray diffraction, and hardness testing. A metallographic examination of the failed bearings should reveal the presence of untempered martensite and its intimate relationship to the initiation and propagation of the cracks. Microhardness testing would reveal that the phase through which the cracks are propagating is indeed sufficiently hard to be untempered martensite. X-ray diffraction can reveal the presence of retained austenite, since the austenite is face-centered cubic, while the preferred martensite has a body-centered structure. The examination of unused bearings could confirm its presence after manufacture. Finally, the presence of retained austenite in manufactured bearings could be further confirmed by the presence of abnormally low bulk hardness values (Austenite is considerably weaker than the desired tempered martensite).

2). Retained austenite can be prevented by modifying the quench to provide for continuous cooling to sub-zero temperatures, either through refrigeration or immersion in liquid nitrogen. By following the quench with a tempering and finish grinding, the bearings could be produced with the desired tempered martensite structure. (NOTE: The quench should not be interrupted by any lengthy holds or delays, as these may lead to a stabilization of the austenite and a suppression of its further transformation.)

APPENDIX K - CASE STUDY
Fire Extinguisher Pressure Gage

1). A number of questions come to mind. How many failures have been recorded? Are they all from the same batch or production run? How long have the failed components been in service? Under what conditions of temperature, humidity, corrosive environment, etc.? Have they been serviced or recharged? If so has the maintenance been performed properly? What gases or chemicals might be present in the interior of the tube? Are these potentially dangerous or might they react with the tubing? What is the normal internal pressure? Could the chemicals present in the extinguisher have played a role? How was the tube manufactured? Was the starting tubing seamless or seamed tubing? How much cold work was imparted to the tubing? Was a stress relief or anneal incorporated after forming? What was the likely ductility of the tubes when put into service? Were the tubes inspected? If so, how? In the failed components, is the failure by a single crack or multiple cracks? Do the cracks have a branching appearance? Do they follow the flow lines of deformation? Are they intergranular or transgranular? Are any corrosion products observable? Is there evidence of any plastic deformation, such as would be present if the tubing had burst? Could mishandling have caused the damage?

2). The tubing could have been defective as it came from the original supplier. If the tubing was seamed tubing, this could be the location of a poor bond. Massive inclusions, seams, laps, and other metallurgical defects could produce failures of this sort. If this were the case, there should be some correlation to tubing supplier, date of manufacture or batch, etc. Also, metallographic examination should reveal features that confirm the presence of metallurgical defects in the tubing. In this case, the cracks should have formed as the bourdon tube was being manufactured. Defects of this type should have been detected by the manufacturer.

An overpressurization could have occurred, causing the tubing to burst. In this case, plastic deformation should be observable and the fractured regions should be flared toward the outside of the tubing. A single burst should be present, and the fracture would most likely be transgranular.

Copper-base alloys are also susceptible to stress-corrosion cracking, especially when present in moist or humid environments. If this were the case, metallography would reveal the crack to be brittle in appearance, following grain boundaries in the direction of prior working, and be a branching crack (most likely, multiple cracks should be present). The absence of a prior anneal or stress relief would be noted. Standard tests could be conducted to determine the susceptibility of the particular material to stress-corrosion cracking.

Mechanical abuse might also be considered, but for an expectedly ductile material, there should be signs of plastic deformation that would have preceded final fracture.

3). Of the mechanisms proposed above, only stress-corrosion-cracking would account for a satisfactory product being made at the manufacturer and the defect forming at a later time when the product is in service. Cracking due to defective tubing should have occurred during the process of forming the bourdon tube. Overpressurization would likely have occurred during either the initial manufacture (failure should have been noted), or during recharging (a correlation of failures and service record should be noted). Mechanical abuse should come with accompanying signs of prior deformation.

4). Assuming that the failure mechanism is indeed stress-corrosion-cracking, possible alternatives would be to subject all formed bourdon tubes to a stress-relief or anneal heat treatment. Elimination of the corrosive environment would be extremely difficult, so the problem should be addressed through the stress approach. Another alternative would be to change the material in the bourdon tube to a metal or alloy with reduced susceptibility to this particular mode of failure.